Modules
Tools for Teaching 1991

published by

Consortium for Mathematics
and Its Applications, Inc.
Suite 210
57 Bedford Street
Lexington, MA 02173

edited by

Paul J. Campbell
Campus Box 194
Beloit College
700 College Street
Beloit, WI 53511–5595
campbell@beloit.edu

All rights reserved. No part of this publication may be reproduced, stored in a retrieval system, or transmitted in any form or by any means, electronic, mechanical, photocopying, recording, or otherwise, without prior permission of the copyright holder.

Recommendations expressed are those of the authors and do not necessarily reflect the views of the copyright holder.

© 1992 COMAP, Inc. All rights reserved.
ISBN 0–912843–22–5
Printed in USA.

Table of Contents

Introduction v

Unit		Page
708	Computer and Calculator Computation of Elementary Functions .... 1 Richard J. Pulskamp and James A. Delaney	
712	The Drag Force on a Sphere .... 35 H. Edward Donley	
714	Heat Therapy for Tumors .... 71 Leah Edelstein-Keshet	
716	Newton's Method and Fractal Patterns .... 103 Philip D. Straffin, Jr.	
717	3D Graphics in Calculus and Linear Algebra .... 125 Yves Nievergelt	
726	Calculus Optimization in Information Technology .... 171 Paul J. Campbell	

Introduction

The instructional modules in this volume were developed by the Undergraduate Mathematics and Its Applications (UMAP) Project. UMAP has been funded by grants from the National Science Foundation to Education Development Center, Inc. (February 1976–February 1983) and to the Consortium for Mathematics and Its Applications (COMAP), Inc. (February 1983–February 1985). Project UMAP develops and disseminates instructional modules and expository monographs in the mathematical sciences and their applications for undergraduate students and instructors.

UMAP Modules are self-contained (except for stated prerequisites), lesson-length, instructional units from which undergraduate students learn professional applications of mathematics and statistics to such fields as biomathematics, economics, American politics, numerical methods, computer science, earth science, social sciences, and psychology. The modules are written and reviewed by classroom instructors in colleges and high schools throughout the United States and abroad. In addition, a number of people from industry are involved in the development of instructional modules.

In addition to the annual collection of UMAP Modules, COMAP also distributes individual UMAP instructional modules, *The UMAP Journal*, and the UMAP expository monograph series. Thousands of instructors and students have shared their reactions to the use of these instructional methods in the classroom. Comments and suggestions for changes are incorporated as part of the development and improvement of materials.

The substance and momentum of the UMAP Project comes from the thousands of individuals involved in the development and use of UMAP's instructional materials. In order to capture this momentum and succeed beyond the period of federal funding, we established COMAP as a nonprofit organization. COMAP is committed to the improvement of mathematics education, to the continuation of the development and dissemination of instructional materials, and to fostering and enlarging the network of people involved in the development and use of materials. COMAP deals with science and mathematics education in secondary schools, teacher training, continuing education, and industrial and government training programs.

Incorporated in 1980, COMAP is governed by a Board of Trustees:

David Roselle, President	University of Delaware
Robert M. Thrall, Treasurer	Rice University, Retired
William F. Lucas, Clerk	Claremont Graduate School
Linda M. Chaput	W.H. Freeman & Co.
Margaret B. Cozzens	Northeastern University
H. Newton Garber	Garber Associates, Inc.
Landy Godbold	The Westminster Schools
Marion T. Jones	Devitt-Jones Productions, Ltd.
Trudi C. Miller	Univ. of Minnesota–Minneapolis
Jewell Plummer–Cobb	Cal. State Univ.–Fullerton
Henry Pollak	Bell Comm. Research, Retired

Instructional programs are guided by the Consortium Council, whose members are elected by the broad COMAP membership, or appointed by cooperating organizations (Mathematical Association of America, Society for Industrial and Applied Mathematics, National Council of Teachers of Mathematics, American Mathematical Association of Two-Year Colleges, The Institute of Management Sciences, and American Statistical Association). The 1991 Consortium Council is chaired by Margaret Barry Cozzens (National Science Foundation), and its members are:

Ronald Barnes	University of Houston–Downtown
Donna Beers	Simmons College
Richard G. Brown	Phillips Exeter Academy
Alphonse Buccino	University of Georgia–Athens
Paul J. Campbell	Beloit College
Toni Carroll	Siena Heights College
Margaret B. Cozzens	Northeastern University
Gary Froelich	Bismarck High School
Bernard Fusaro	Salisbury State University
Frank R. Giordano	U.S. Military Academy
Landy Godbold	The Westminster Schools
Charles Hamberg	Illinois Math & Science Acad.
John Kenelly	Clemson University
Peter Lindstrom	North Lake College
Kay Merseth	Univ. of California–Riverside
Walter Meyer	Adelphi University
Fred S. Roberts	Rutgers University
Stephen Rodi	Austin Community College
Gene Woolsey	Colorado School of Mines

This collection of modules represents the spirit and ability of scores of volunteer authors, reviewers, and field-testers (both instructors and students). The modules also present various fields of applications as well as different levels of mathematics. COMAP is very interested in receiving information on the use of modules in various settings. We invite you to contact us.

UMAP

Modules in Undergraduate Mathematics and its Applications

Published in cooperation with the Society for Industrial and Applied Mathematics, the Mathematical Association of America, the National Council of Teachers of Mathematics, the American Mathematical Association of Two-Year Colleges, The Institute of Management Sciences, and the American Statistical Association.

Module 708

Computer and Calculator Computation of Elementary Functions

Richard J. Pulskamp
James A. Delaney

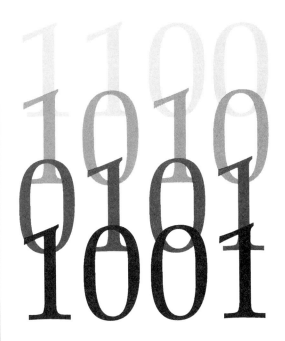

Applications of Numerical Analysis and Calculus to Computation

COMAP, Inc., Suite 210, 57 Bedford Street, Lexington, MA 02173 (617) 862–7878

INTERMODULAR DESCRIPTION SHEET:	UMAP Unit 708
TITLE:	Computer and Calculator Computation of Elementary Functions
AUTHOR:	Richard J. Pulskamp James A. Delaney Dept. of Mathematics and Computer Science Xavier University 3800 Victory Parkway Cincinnati, OH 45207
MATHEMATICAL FIELD:	Numerical analysis, Calculus
APPLICATION FIELD:	Computation
TARGET AUDIENCE:	Second-semester calculus students, computer science students.
ABSTRACT:	We consider methods used to approximate the elementary functions on digital computers and electronic calculators. The requirements for such approximations are discussed, with brief comments on hardware and error. We emphasize range reduction, polynomial approximations, and especially CORDIC techniques. We treat the square root, trigonometric, exponential, and logarithmic functions in detail.
PREREQUISITES:	Calculus through Taylor's formula. Some of the exercises assume access to a graphing calculator or a computer graphing program. We recommend use of a calculator and a computer in connection with this Module.

©Copyright 1991, 1992 by COMAP, Inc. All rights reserved.

COMAP, Inc., Suite 210, 57 Bedford Street, Lexington, MA 02173
(800) 77–COMAP = (800) 772–6627,   (617) 862–7878

Computer and Calculator Computation of Elementary Functions

Richard J. Pulskamp
James A. Delaney
Dept. of Mathematics and Computer Science
Xavier University
3800 Victory Parkway
Cincinnati, OH 45207

Table of Contents

1. INTRODUCTION	1
2. BACKGROUND INFORMATION	3
2.1 Hardware Aspects	3
2.1.1 The base used for number representation	3
2.1.2 Integer vs. real-number representation and arithmetic	4
2.1.3 Hardware vs. software implementation of operations	4
2.2 Requirements of Function Evaluation Routines	5
2.3 Error Considerations	6
3. THE SQUARE ROOT	7
3.1 Calculator Computation of Square Root	7
3.2 Computer Computation of the Square Root	9
4. ELEMENTARY FUNCTIONS ON COMPUTERS	10
4.1 Preliminary Remarks	10
4.2 Computation of Elementary Functions	12
4.2.1 Computation of the exponential function	12
4.2.2 Evaluation of the natural logarithm	14
5. ELEMENTARY FUNCTIONS ON CALCULATORS	15
5.1 Sines, Cosines, and Tangents via CORDIC	15
5.1.1 The CORDIC algorithm for tangent	17
5.1.2 The CORDIC computation of sines and cosines	21
5.2 A CORDIC Algorithm for Arctangent	21
5.2.1 Derivation of the CORDIC algorithm for arctangent	23
6. SOLUTIONS TO THE EXERCISES	25
REFERENCES	28
ACKNOWLEDGMENTS	30
ABOUT THE AUTHORS	30

MODULES AND MONOGRAPHS IN UNDERGRADUATE
MATHEMATICS AND ITS APPLICATIONS (UMAP) PROJECT

The goal of UMAP is to develop, through a community of users and developers, a system of instructional modules in undergraduate mathematics and its applications, to supplement existing courses and from which complete courses may be built.

The Project was initially funded by a grant from the National Science Foundation and has been guided by a National Advisory Board of mathematicians, scientists, and educators. UMAP is now supported by the Consortium for Mathematics and Its Applications (COMAP), Inc., a non-profit corporation engaged in research and development in mathematics education.

COMAP Staff

Paul J. Campbell	Editor
Solomon Garfunkel	Executive Director, COMAP
Laurie W. Aragón	Development Director
Philip A. McGaw	Production Manager
Roland Cheyney	Project Manager
Laurie M. Holbrook	Copy Editor
Dale Horn	Design Assistant
Rob Altomonte	Distribution Coordinator
Sharon McNulty	Executive Assistant

1. Introduction

We routinely tackle computational problems with the aid of calculators or by programming a computer. In the process, we assume the availability of rapid and accurate evaluation of such elementary functions as $\sqrt{x}$, $\exp x$, $\cos x$, and $\arctan x$.

It may come as a surprise that sometimes the values reported by a machine are inaccurate. For example, BASICA on an IBM PC returns $\ln 1.001$ as 9.994461E−4 instead of the correct value 9.995003E−4; only the first three digits are correct! Yet, the little Casio fx–7000G graphing calculator gives 9.995003331E−4 with all ten digits correct. Obviously, not everyone is doing the same thing.

How the elementary functions can be calculated accurately, rather than having inaccuracies in some existing routines, is the main focus of this Module. In truth, the "obvious" way to compute the value of a function is usually not the best way. As a result, computer and calculator manufacturers have implemented many interesting algorithms for this purpose.

To illustrate how calculations are really performed, we will describe some algorithms in current use. For computer routines, we draw upon the run-time library of the Digital Equipment Corporation VAX series of computers. For calculators, we discuss routines used by Hewlett-Packard in its calculators.

The choice of method for function evaluation is dependent upon many factors. The most important of these is what we might call the hardware constraint. While hardware details are examined in greater detail in Section 2.1, a few remarks are in order now.

Consider first typical calculators. These machines are designed to perform a very limited collection of operations. Algorithms to perform arithmetic and to compute the elementary functions must be closely matched with the electronic components. In a calculator, the transcendental functions are computed using the little-known method called *decimal CORDIC*. "CORDIC" stands for COordinate Rotation DIgital Computer. It is an ingenious method that uses formulas for functions of a sum as a basis for a sequence of recursive computations that converge to the desired function value. The CORDIC procedure has the advantage that a very limited instruction set will permit the evaluation of all of the elementary functions. We will discuss CORDIC at length later in Section 5. To the best of our knowledge, all present-day calculators use a CORDIC method to compute values of the elementary functions.

Computers, on the other hand, being general purpose machines, allow more flexibility in the choice of algorithm. Aside from the CORDIC method, which is used in numeric coprocessors of personal computers, three general classes of procedures are available to compute the elementary functions. These procedures are

- analytic function approximation, such as Taylor series;
- iterative techniques, such as the Newton-Raphson method; or
- table lookup coupled with a means of interpolation.

We close this introduction with a mention of Taylor and Maclaurin series. These series are not universally used in computers, because often there are better approximations. But Taylor series do illustrate some general considerations for all function approximation. Among these are

- computability with only elementary arithmetic operations;
- error estimates, to provide information about accuracy; and
- range reduction.

The reader is encouraged to work the exercises as they are encountered in the text. Some exercises ask that errors be estimated or routines be compared. This can be done most easily by using a function-graphing program (with double precision, if possible) on a computer. When doing this, assume that the computer is able to produce exact values for the elementary functions.

Exercises

1. Determine how many terms of the Maclaurin series for e^x are needed to estimate e^x to six-digit accuracy
   a) on the interval $[-1, 1]$;
   b) near $x = 10$.

2. Investigate the absolute error that results in using the first five terms of the Maclaurin series to approximate e^x on the interval $[-1, 1]$. This can be done by graphing

$$y = e^x - \left(1 + x + \frac{x^2}{2} + \frac{x^3}{6} + \frac{x^4}{24}\right).$$

Where is the error maximal?

2. Background Information

To understand what goes into the design of an algorithm, the reader must be aware of hardware aspects, general function evaluation requirements, and error considerations. A discussion of these follows.

2.1 Hardware Aspects

We need at least a rudimentary understanding of a few hardware concepts, including:

- the base used for number representation,
- integer vs. real-number representation and arithmetic, and
- hardware vs. software implementation of arithmetic operations.

2.1.1 The base used for number representation

Physically, numbers are represented through the use of binary (two-state) elements. Whether these are considered on-off, or up-down, etc., we describe them in terms of the two digits 0 and 1. Thus the most natural base to use for integer storage is two. We can use four bits to represent the integers from 0000 to 1111 binary, equivalent to 0 to 15 decimal. Longer sequences of binary digits will represent larger integers. One byte (eight bits) can represent unsigned integers from 0 to 255, or signed integers from -128 to 127. Frequently, we group the binary digits four at a time and describe them as hexadecimal digits. For example,

$$0000 \text{ binary} = 00 \text{ decimal} = 0 \text{ hex,}$$
$$1010 \text{ binary} = 10 \text{ decimal} = A \text{ hex,}$$
$$1111 \text{ binary} = 15 \text{ decimal} = F \text{ hex.}$$

Thus one byte is usually treated as two hex digits. This is an almost universal method adopted for general-purpose digital computers.

Another representation scheme, BCD (*Binary Coded Decimal*), retains base ten. This system is employed in most calculators, which use only the binary configurations 0000 through 1001 to represent the decimal digits 0 through 9. By way of comparison, note that the integer twelve has the bit representation 0001 0010 in BCD, but the representation 0000 1100 in binary (i.e., 0C hex.)

Integers are represented exactly in any base; but the same is not necessarily true for other numbers, because a machine can store only finitely many digits. Whether or not these numbers have exact representations depends on the base and on the number of digits retained. In base ten, the number one-tenth has the exact representation 0.1000.... Whether our device rounds or truncates, we lose only zeros. By contrast, the number two-thirds in decimal is 0.6666..., repeating infinitely. Whether we round or truncate, the representation is not exactly two-thirds.

In binary, one-half has the representation 0.1000..., and one-quarter has the representation 0.0100... exactly. But in binary, one-tenth is an infinite repeating nonzero sequence of binary digits following the binary point. No matter where the approximation is terminated, one-tenth cannot be represented exactly in any binary (or essentially binary, such as hexadecimal) number system. This fact is not always obvious from a computer printout. If a machine reads 0.1 and prints it (with a default format), the printout usually shows 0.1. But if on a VAX we specify printing of ten decimal places, the printed value is 0.1000000015.

Another important consequence of the base occurs in multiplication and division. In base ten, multiplications and divisions by a power of ten can be accomplished by simple shifting of the decimal point. Multiplications by 2, 4, 8, 16, etc., are more cumbersome. In binary, on the other hand, multiplications and divisions by powers of 2 can be accomplished by shifting the binary point, whereas multiplications by 10 are more cumbersome.

2.1.2 Integer vs. real-number representation and arithmetic

Two very common computer number representations are *fixed-point* (integer) and *floating point* (real). Fixed-point corresponds to the familiar representation (without a decimal point) of integer values. Floating-point closely resembles scientific notation, involving two components, the *mantissa* (significant digits) and the *characteristic* (exponent on the floating point base.) Each representation has its own set of arithmetic algorithms. Fixed-point arithmetic is comparatively simple and quick, while floating-point arithmetic is not, because operations must be performed on both mantissa and characteristic.

Exercise

3. Add 0.23×10^5 and 0.53×10^{-4} while retaining scientific notation.

2.1.3 Hardware vs. software implementation of operations

Basic arithmetic operations, such as addition and multiplication, are not themselves among the most fundamental hardware operations. Among the most primitive operations, from which more advanced ones can be built, are bit-by-bit AND, OR, and NOT. With these as building blocks, shifts, compares (as in IF statements), etc., are themselves primitive, hence fast. More advanced still, built up from primitive operations, are bit-by-bit adders with carry, multibit adders, etc., and sometimes multiplication, division, and even square root.

The designers of a calculator or computer decide on the level of complexity that they want to build into the hardware as machine instructions. The instruction set might be quite limited, as for most calculators, or very extensive. The VAX, in fact, has a single machine instruction to evaluate an entire polynomial. The more operations that are built into the hardware, the more complicated and hence more expensive the processor. The operations necessary for problem solution then must be either built into the hardware or programmed as a sequence of built-in instructions (software or firmware.)

In summary, the suitability of a mathematical algorithm for a particular computing device depends in large part on the operations that are built in as machine instructions and therefore are fast. Operations that must be accomplished with programming may be relatively—and sometimes prohibitively—slow.

2.2 Requirements of Function Evaluation Routines

Let's introduce some notation. Denote by f the function to be evaluated and by F the routine that is used to approximate f computationally. Let x denote an element of the domain of f. The list below, adapted from Fike [1968], gives requirements for any algorithm acceptable for general use.

- **Accuracy.** Ideally, for each x, the value $f(x)$ should equal $F(x)$ to the precision of the computations.

- **Speed.** The time required to compute the value of a function is dominated by the number of multiplications (and divisions) performed. The fastest algorithms use only addition, shifting, and comparisons.

- **Special arguments.** We expect the value of a function for special arguments to be exact to the limit of machine precision. For example, if $x = 0$, then $\cos x$ should be reported as exactly 1.

- **Invalid arguments.** There are two aspects to consider. First, a function itself may not be defined for every real number, as, for example, $\sqrt{x}$ is not defined for $x < 0$. The algorithm must test for such values. Second, every machine imposes limits on the magnitudes it can represent. This reduces the domain over which a function is defined computationally.

- **Bounds on F.** When f is bounded, then F must be as well. An obvious example is $\sin x$ or $\cos x$. No approximation of these functions should ever exceed 1 for any value of x.

- **Monotonicity.** If f is an increasing or decreasing function, then so must F be increasing or decreasing.

2.3 Error Considerations

Of paramount importance is the maintenance of an acceptable level of accuracy over the entire domain of the function. In the case of a Maclaurin series, a few terms are usually sufficient to achieve great accuracy if x is near 0. However, as $|x|$ increases, accuracy can only be achieved by using an ever-increasing number of terms of the series. In theory, we could overcome this problem by using different Taylor series representations, each restricted to a particular subinterval of the domain. But this is not a realistic solution, because most domains are quite extensive.

A far superior method, *reduction of range*, is used in practice. The general idea is to find an algorithm for the function which is very accurate and efficient on a comparatively small interval, the *fundamental interval*. For the method to be applicable to values outside of this interval, a transformation is needed to express $f(x)$ for arbitrary x in terms of $f(x)$ in the primary interval. This transformation and its inverse transformation must be easy to compute.

Each function algorithm has its own specific range-reduction procedure. One simple example to illustrate the process is $f(x) = \log_{10} x$. Write x as $z \times 10^n$, where $1 \leq z < 10$ and n is an integer. Then

$$\log_{10} x = \log_{10}(z \times 10^n) = \log_{10} z + \log_{10} 10^n = \log_{10} z + n.$$

Now all that is needed is a very accurate formula for $\log_{10} z$ for $1 \leq z < 10$.

The problem of error really reduces to just one question: How many significant digits will the procedure produce? While it is not the purpose of this Module to discuss error analysis in depth, we need to discuss sources of error.

Suppose we wish to evaluate the function f at the value x, where x itself is exact. (An error in x itself would complicate the error analysis even further.) We approximate f by the formula or algorithm F. The formula F usually cannot be used directly but rather is implemented on a machine. Let's call the implemented formula F^*. Thus, we see that any procedure will involve the three possibly distinct values $f(x)$, $F(x)$, and $F^*(x)$. There are three different error quantities of interest:

- the *truncation error* $\epsilon_1(x) = |f(x) - F(x)|$. The terminology arises from the Taylor series representations of the elementary functions; we may always take F to be the truncated series.

- the *rounding error* $\epsilon_2(x) = |F(x) - F^*(x)|$. Rounding error is the result of the precision with which the computations are performed, the order of operations, whether intermediate values are rounded or truncated, and peculiarities of machine design.

- the *absolute error* $\epsilon(x) = |f(x) - F^*(x)|$. We have $\epsilon(x) \leq \epsilon_1(x) + \epsilon_2(x)$.

Generally, it is possible to obtain a good estimate of the truncation error. As a result, we are able to design formulas that are highly accurate. More difficult to estimate (and to control) is the rounding error and especially how it propagates throughout a formula.

Another source of error (which can arise from the conversion of decimal to binary) is *conditioning* or *sensitivity*. Conditioning measures the effect on the function of an inaccuracy in the argument. Sometimes a small error in an independent variable produces a small error in the dependent variable. However, it may also happen that a small error in the independent variable can produce an enormous error in the dependent variable, over and above any computational errors. Such is the case for $\ln x$ near 1. A small inaccuracy in x, due, for example, to conversion of exact decimal to approximate binary, can produce large relative error in the corresponding logarithm. A VAX 11/785 computes $\ln 1.001$ as $9.99547E-4$. The last two digits are in error. However, $\ln(1025/1024)$ is computed correctly as $9.76086E-4$. Why the difference? The number 1.001 does not have a terminating binary representation, but 1025/1024 does.

3. The Square Root

There are several reasons why we discuss the computation of the square root in detail. The square root is used within other algorithms, such as those involving trigonometric identities. The calculator version (Section 3.1) of the computation shows some of the features that distinguish the CORDIC from analytic procedures. The computer version (Section 3.2) is the only example of an iterative analytic procedure in this Module. Both procedures are short and simple.

3.1 Calculator Computation of Square Root

In order to compute $\sqrt{x}$, let us first write x in scientific notation so that $x = z \times 10^k$. We may always choose k even and $1 \leq z < 100$. Since $\sqrt{x} = \sqrt{z} \times 10^{k/2}$, we need only compute $\sqrt{z}$.

The reader must keep in mind that the algorithm must be one that is easily implemented in hardware. The key to the computation is the fact that the sum of the first n odd numbers is precisely n^2: for example, $1 + 3 + 5 + 7 + 9 = 5^2$.

The procedure that we describe is essentially that used by Hewlett-Packard in its calculators (see [Egbert 1977a]). By design, the square root is computed from below, digit by digit. Let a_1 be the one-digit approximation to $\sqrt{z}$; and, in general, let a_j be the j-digit approximation. Given a_j, we wish to compute the next digit b in the approximation. Notice that a_1 is the units digit (since $1 \leq a_1 < 10$). Therefore, $a_{j+1} = a_j + b \times 10^{-j}$, for $j = 1, 2, 3, \ldots$.

Let $R_j = z - a_j^2$; R_j is a measure of the "goodness" of a_j as $\sqrt{z}$. By approaching z from below, we have $R_j \geq 0$ for all j, with

$$R_{j+1} = z - a_{j+1}^2 = z - \left(a_j + b \times 10^{-j}\right)^2$$

$$\begin{aligned}
&= z - \left(a_j{}^2 + 2a_j b \times 10^{-j} + b^2 \times 10^{-2j}\right) \\
&= z - a_j{}^2 - \left(2a_j b \times 10^{-j} + b^2 \times 10^{-2j}\right) \\
&= R_j - \left(2a_j b \times 10^{-j} + b^2 \times 10^{-2j}\right).
\end{aligned}$$

The strategy is to choose b so as to make R_{j+1} a minimum. The form above is not optimal for implementation on a calculator, because the term containing a_j has other factors. To improve it, first multiply through by 5 and then eliminate the factor of a power of 10 from the term $a_j b$. This yields

$$10^{j-1} \times 5R_{j+1} = 10^{j-1} \times 5R_j - \left(a_j b + 5b^2 \times 10^{-j-1}\right).$$

Using the fact that $b^2 = \sum_{i=1}^{b}(2i-1)$ and writing $a_j b$ as $\sum_{i=1}^{b} a_j$, we obtain

$$10^{j-1} \times 5R_{j+1} = 10^{j-1} \times 5R_j - \sum_{i=1}^{b}\left(a_j + 5(2i-1) \times 10^{-j-1}\right).$$

Since $5(2i-1) = 05, 15, 25, \ldots, 95$ as $i = 1, 2, 3, \ldots, 10$, we may represent $5(2i-1)$ as $(i-1)|5$ and write

$$10^{j-1} \times 5R_{j+1} = 10^{j-1} \times 5R_j - \sum_{i=1}^{b}\left(a_j + (i-1)|5 \times 10^{-j-1}\right). \quad (1)$$

We may initialize the process with $a_0 = 0$ and $R_0 = z$ in (1) (see **Exercise 4**). Indeed, this is what is done in the calculator. By hand, it is easier for us to find a_1 and R_1 directly. After this point, (1) can be used repeatedly. We must emphasize that it is not necessary to know R_j itself; the calculation requires only that we know $10^{j-1} \times 5R_j$.

Example. Compute $\sqrt{27.17954}$.

$R_1 = 27.17954 - (1 + 3 + 5 + 7 + 9) = 2.17954$. Clearly, $a_1 = 5$.
To find a_2, we put $j = 1$ in (1). This gives

$$\begin{aligned}
5R_2 &= 5R_1 - \sum_{i=1}^{b}\left(a_1 + (i-1)|5 \times 10^{-2}\right) \\
&= 10.8977 - \sum_{i=1}^{b}\left(5 + (i-1)|5 \times 10^{-2}\right) \\
&= 10.8977 - (5.05 + 5.15) \\
&= 0.6977.
\end{aligned}$$

Therefore, $b = 2$ (since we subtracted two terms) and $a_2 = 5.2$. This is the only time we use multiplication by 5.
To find a_3, put $j = 2$ in (1). This gives

$$10 \times 5R_3 = 10 \times 5R_2 - \sum_{i=1}^{b}\left(a_2 + (i-1)|5 \times 10^{-3}\right)$$

$$= 6.977 - \sum_{i=1}^{b}\left(5.2 + (i-1)|5 \times 10^{-3}\right)$$
$$= 6.977 - (5.205)$$
$$= 1.772.$$

Therefore, $b = 1$ (one subtraction) and $a_3 = 5.21$.

To find a_4, put $j = 3$ in **(1)**. This gives

$$10^2 \times 5R_4 = 10^2 \times 5R_3 - \sum_{i=1}^{b}\left(a_3 + (i-1)|5 \times 10^{-4}\right)$$
$$= 17.72 - \sum_{i=1}^{b}\left(5.21 + (i-1)|5 \times 10^{-4}\right)$$
$$= 17.72 - (5.2105 + 5.2115 + 5.2125)$$
$$= 2.0855.$$

Therefore, $b = 3$ and $a_4 = 5.213$.

To find a_5, put $j = 4$ in **(1)**. This gives

$$10^3 \times 5R_5 = 10^3 \times 5R_4 - \sum_{i=1}^{b}\left(a_4 + (i-1)|5 \times 10^{-5}\right)$$
$$= 20.855 - \sum_{i=1}^{b}\left(5.213 + (i-1)|5 \times 10^{-5}\right)$$
$$= 20.855 - (5.21305 + 5.21315 + 5.21325 + 5.21335)$$
$$= 0.0022.$$

Therefore, $b = 4$ and $a_5 = 5.2134$. We will stop here.

Note the simplicity of the operations involved. With the exception of the multiplication by 5 in step 1, which could be accomplished with additions, the only operations are subtractions, additions, shifts, and comparisons.

Exercise

4. Using the procedure of this section, compute $\sqrt{27.17954}$, beginning with $a_0 = 0$ and $R_0 = z$.

3.2 Computer Computation of the Square Root

Our goal is to compute $\sqrt{x}$ when x is a positive number. One strategy employed is the iterative scheme known as the Newton-Raphson procedure, often called *Newton's method*. A reader who is unfamiliar with this method should refer to any calculus textbook; a reader who is familiar with the method should verify that **(2)** below is the correct instance for $\sqrt{x}$ of the general procedure.

Assume that the computer is a binary machine. A binary-point representation of x is $x = z \times 2^k$, where $\frac{1}{2} \le z < 1$ and k is an integer. For the moment, assume that k is even. Then $\sqrt{x} = \sqrt{z} \times 2^{k/2}$. We need only compute $\sqrt{z}$ and accomplish the multiplication by $2^{k/2}$ via shifting. The procedure we describe is that used in the VAX system library (see [VAX 1988]).

Given the initial approximation a_0 to $\sqrt{z}$, we form the sequence of estimates $a_1, a_2, \ldots$ by the Newton-Raphson recursion

$$a_{i+1} = \frac{1}{2}\left(a_i + \frac{z}{a_i}\right). \tag{2}$$

All that remains is the determination of the initial value a_0. This is done by approximating $\sqrt{x}$ on the interval $\left[\frac{1}{2}, 1\right]$ by a line segment. The segment, chosen to minimize the error at the midpoint and endpoints, is on the line given by the equation $L(x) = 0.5857864x + 0.4204951$. We take $a_0 = L(z)$.

If k is odd, rewrite x as $x = \frac{z}{2} \times 2^{k+1}$. We now must compute $\sqrt{\frac{z}{2}}$, where $\frac{1}{4} \le \frac{z}{2} < \frac{1}{2}$. Again, we construct a sequence of estimates $a_1, a_2, \ldots$ using the Newton-Raphson procedure (2) and starting with $a_0 = 0.8284271\left(\frac{z}{2}\right) + 0.3012412$. This value of a_0 is obtained from a linear approximation to $\sqrt{x}$ on the interval $\left[\frac{1}{4}, \frac{1}{2}\right]$.

The sequence $a_0, a_1, a_2, \ldots$ converges quite rapidly to $\sqrt{z}$. In fact, the VAX needs only two iterations for single-precision accuracy and three for double-precision accuracy.

Exercise

5. Using the procedure described in this section, compute
   a) $\sqrt{0.3}$;
   b) $\sqrt{0.7}$.

4. Elementary Functions on Computers

Most general-purpose computers handle the evaluation of the elementary functions by an explicit function approximation (formula). After some preliminary remarks and examples in Section 4.1, in which various polynomial and rational function approximations to $\exp x$ are briefly discussed, we will exhibit specific methods in Section 4.2 for the calculation of $\exp x$ and $\ln x$.

4.1 Preliminary Remarks

As mentioned earlier, the truncated Taylor series is sometimes a poor choice for the evaluation of a function. It tends to be very accurate only near the point about which it is expanded. There exist polynomials of the same degree, known as *economized polynomials*, which distribute the error more uniformly over an interval.

For $f(x) = e^x$, one economized polynomial for the interval $[-1, 1]$, based upon the Chebyshev polynomials (with coefficients given to 5 decimal places) is

$$C(x) = 1.00005 + 0.99730x + 0.49916x^2 + 0.17736x^3 + 0.04384x^4.$$

We compare this with the truncated Maclaurin series of degree 4,

$$M(x) = 1.00000 + 1.00000x + 0.50000x^2 + 0.16667x^3 + 0.04167x^4.$$

To see the advantage of the economized polynomial, examine the graphs of $E_C(x) = |e^x - C(x)|$ and $E_M(x) = |e^x - M(x)|$ in **Figure 1**.

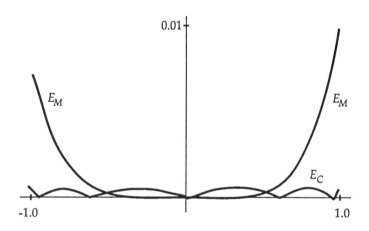

Figure 1. E_C compared to E_M.

An improvement over the Taylor polynomial of degree n can sometimes be achieved by approximating the function by a rational function $R(x) = p(x)/q(x)$, where the degree of p and the degree of q sum to n. One frequently employed rational approximation is the *Padé approximation*, which has an error term different from that of the Taylor polynomial. Again using the exponential function as an example, the Padé approximation $R(x)$ obtained by expanding about the origin is

$$R(x) = \frac{1 + x/2 + x^2/12}{1 - x/2 + x^2/12}.$$

To see the improvement over the corresponding fourth-degree Taylor polynomial, refer to **Figure 2**, which compares the graphs of E_M and $E_R(x) = |e^x - R(x)|$.

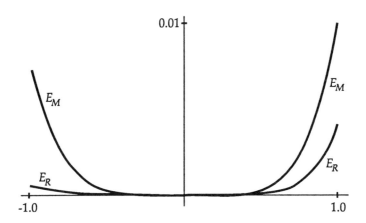

Figure 2. E_R compared to E_M.

For a treatment of Chebyshev polynomials and Padé approximations, see Gerald and Wheatley [1989]. Although in general both the economized polynomial and the Padé approximation are often improvements over the Taylor polynomial, still further improvements can be made. Since some aspects of function approximation remain an art, many different methods are used in practice.

4.2 Computation of Elementary Functions

The first step in the approximation of a function f is to exploit one or more of its properties to achieve a reduction in range. Under such a transformation, every element x in the domain of the function is mapped to a corresponding element z in a small interval, the fundamental interval. The second step is to construct on the fundamental interval a polynomial or rational function $p(z)$, the fundamental approximation formula, which approximates the function f with a high degree of accuracy. To obtain the desired value of f, it is necessary that there be a simple relationship between $f(x)$ and $f(z)$, so that the reduction of range transformation can be easily reversed.

4.2.1 Computation of the exponential function

Since $e^x = 2^{x \log_2 e}$, it suffices to develop a procedure for the computation of the latter—a task well-suited to a binary computer.

Let's examine in detail how the VAX computes e^x in ordinary single-precision floating-point (see [VAX 1988]). First, tests are made for three exceptional cases:

- If $x > 88$, the machine signals overflow.

- If $x \leq -89$, the machine returns $e^x = 0$.
- If $|x| < 2^{-25}$, the machine returns $e^x = 1$.

Otherwise, put $y = x \log_2 e$. Consider y in hexadecimal notation with three components:

- $u =$ the integer part,
- $v =$ the hex digit in the first position right of the hexadecimal point, and
- $z =$ the remaining fractional part.

For example, if $y = 41.\text{B2C}$, then $u = 41$, $v = \text{B}$, and $z = .02\text{C}$. Since $y = u + v/16 + z$, it follows that $2^y = 2^u \times 2^{v/16} \times 2^z$. The crucial facts to observe are

- multiplication by 2^u is achieved by a shift of the binary point or by adjusting the characteristic of the floating point representation;
- the value of $2^{v/16}$ can be found by an easy table lookup (there are only 31 possible values); and
- $\frac{-1}{16} < z < \frac{1}{16}$, thereby making a polynomial approximation to 2^z quite feasible. In fact, the Maclaurin polynomial of degree 4 has a maximum error on this interval of order 10^{-9}.

This procedure requires that $\log_2 e \, (= 1/\ln 2)$ and $\ln 2$ be known with full precision. Historical methods are available for computing isolated values of particular constants such as these. For example, the series

$$\frac{3}{4} - \sum_{n=1}^{\infty} \frac{1}{2n(4n^2-1)^2}$$

converges quite rapidly to $\ln 2$. A standard reference [Abramowitz and Stegun 1965] lists values up to 25 significant digits. These can be stored in memory and made available to any routine that requires them.

Exercise

6. Find the Maclaurin polynomial $p(x)$ of degree 4 for 2^x. Verify that

$$|2^x - p(x)| < 3 \times 10^{-9}$$

on $\left[\frac{-1}{16}, \frac{1}{16}\right]$.

4.2.2 Evaluation of the natural logarithm

For any $x > 0$, x may be written as $2^n \times z$, where $\frac{1}{2} \leq z < 1$. It follows that $\ln x = n \ln 2 + \ln z$. One should note that the Maclaurin series

$$\ln(1+x) = x - \frac{x^2}{2} + \frac{x^3}{3} - \cdots, \qquad |x| < 1, \tag{3}$$

converges quite rapidly in a neighborhood of 0. An efficient algorithm can be based upon this series.

To understand how the computations are carried out, we first make some preliminary observations. The details are relegated to **Exercise 7**.

It is easy to show that

$$\ln\left(\frac{1+x}{1-x}\right) = 2\left(x + \frac{x^3}{3} + \frac{x^5}{5} + \cdots\right), \qquad \text{for } |x| < 1. \tag{4}$$

Consequently, if $X > 0$, there are choices of $A > 0$ so that $V = (X-A)/(X+A)$ satisfies $|V| < 1$. By direct substitution into (4), we obtain

$$\ln\left(\frac{X}{A}\right) = 2\left(V + \frac{V^3}{3} + \frac{V^5}{5} + \cdots\right), \qquad \text{for } |V| < 1,$$

or equivalently,

$$\ln X = \ln A + 2V\left(1 + \frac{V^2}{3} + \frac{(V^2)^2}{5} + \cdots\right). \tag{5}$$

In practice, we choose A in such a way that $|V| \ll 1$. The values of A and $\ln A$ can be stored in a table. When x is near 1, a very short table can contain enough values to assure that V is close to 0. With these ideas in mind, the computation of $\ln x$ is quite straightforward.

We now describe the strategy employed by the VAX routines (see [VAX 1988]). Let us first define the variables N and Z. Write $x = Z \times 2^N$, where $1 \leq Z < 2$ if $x \geq 1$ and $\frac{1}{2} \leq Z < 1$ if $x < 1$. If $|Z-1| < 2^{-5}$, then let $W = Z - 1$ and use the polynomial of degree 4 obtained from equation (3) to compute $\ln Z = \ln(1+W)$. Otherwise, use a table lookup to find the value of A with which to set $V = (Z - A)/(Z + A)$. The polynomial in V of degree 5 derived from (5) is evaluated.

Exercises

7. This exercise develops the approximation of $\ln x$.

    a) Prove that

    $$\ln\left(\frac{1+x}{1-x}\right) = 2\left(x + \frac{x^3}{3} + \frac{x^5}{5} + \cdots\right), \qquad \text{for } |x| < 1.$$

    b) For $V = (X - A)/(X + A)$, derive (5). (Hint: Replace x by V in (4).)

8. Compute
   a) $\ln 1.001$
   b) $\ln 0.78$ using $A = 0.75$ and $\ln A = -0.28768$.

5. Elementary Functions on Calculators

Computations of elementary functions on an electronic calculator are usually performed by some variant of a decimal CORDIC algorithm. This technique was first employed to compute the arctangent and simultaneously the sine and the cosine (see [Volder 1959]). The only machine operations required are shifts, adds, subtracts, compares, and references to stored constants. Thus, CORDIC is ideally suited for implementation on an electronic calculator. We will examine the CORDIC algorithms for sine, cosine, and arctangent only; and we assume that all angles are rendered in radians, not in degrees. Similar algorithms have been developed for the exponential function, the natural logarithm, and, in fact, all the elementary functions.

We will not repeat here the details of range reduction given earlier in the module but assume such a reduction as needed by individual algorithms.

5.1 Sines, Cosines, and Tangents via CORDIC

We begin by investigating a CORDIC algorithm for $\tan x$, where we assume that $x > 0$. A first step is to write x as a finite sum of very special angles, usually the angles whose tangents are the T's in **Table 1**. A larger table would be used in practice, with more decimal places.

Table 1.
Arctangents of powers of 10.

T	1	0.1	0.01
$\arctan T$	0.785398	0.099669	0.010000

The decomposition of x is obtained by successively subtracting multiples of $\arctan 1$, $\arctan 0.1$, $\arctan 0.01$, etc., until the remaining angle a_0 is very small, and while the reduced angles remain nonnegative. With our choice of T's, we have $a_0 < 0.01$. A record must be maintained of the T's used in the decomposition, and of a_0. The examples below illustrate that the decomposition of x is accomplished computationally by simple table lookups followed by subtraction.

Example. The decompositions of $x = 0.99474$, $x = 3$, and $x = 0.5$ are:

$$
\begin{aligned}
0.99474 &= 1\arctan 1 + 2\arctan 0.1 + 1\arctan 0.01, \\
3 &= 3\arctan 1 + 6\arctan 0.1 + 4\arctan 0.01 + 0.005792, \\
0.5 &= 0\arctan 1 + 5\arctan 0.1 + 0\arctan 0.01 + 0.001655.
\end{aligned}
$$

We outline the CORDIC procedure in **Algorithm 1** and illustrate it with an easy example.

Algorithm 1.
The CORDIC algorithm for $\tan x$.

$$\tan x = P_n/Q_n$$

where $\quad x = a_0 + \sum_{i=1}^{n} \arctan T_i \qquad T_i \in \{1, 0.1, 0.01, \ldots\}$

$$P_0 = a_0 \qquad Q_0 = 1$$

$$P_i = P_{i-1} + T_i Q_{i-1} \qquad Q_i = Q_{i-1} - T_i P_{i-1}$$

Example. Compute $\tan 0.99474$.

From the example above, we have

$$0.99474 = 1 \arctan 1 + 2 \arctan 0.1 + 1 \arctan 0.01.$$

Hence we may take $T_1 = 1$, $T_2 = T_3 = 0.1$, $T_4 = 0.01$, and $a_0 = 0$. Using the recurrence relations, we have the results in **Table 2**.

Table 2.
The CORDIC algorithm for $\tan 0.99474$.

	$P_0 = 0.00000$	$Q_0 = +1.00000$
$T_1 = 1$	$+T_1 Q_0 = 1.00000$ $P_1 = 1.00000$	$-T_1 P_0 = -0.00000$ $Q_1 = +1.00000$
$T_2 = 0.1$	$+T_2 Q_1 = 0.10000$ $P_2 = 1.10000$	$-T_2 P_1 = -0.10000$ $Q_2 = +0.90000$
$T_3 = 0.1$	$+T_3 Q_2 = 0.09000$ $P_3 = 1.19000$	$-T_3 P_2 = -0.11000$ $Q_3 = +0.79000$
$T_4 = 0.01$	$+T_4 Q_3 = 0.00790$ $P_4 = 1.19790$	$-T_4 P_3 = -0.01190$ $Q_4 = +0.77810$

Notice that the multiplication of P or Q times T is simple shifting. We find that $\tan 0.99474 = P_4/Q_4 = 1.1979/0.7781 = 1.53952$.

The details of the derivation of this algorithm follow in Section 5.1.1. The reader may, however, proceed directly to Section 5.1.2 to see how CORDIC is used to compute sines and cosines.

Computer and Calculator Computation of Elementary Functions   21

5.1.1   The CORDIC algorithm for tangent

The idea behind the CORDIC method for computing $\tan x$ is that $\sin x$ and $\cos x$ each have a "workable" formula for the sum of two angles, namely,

$$\begin{aligned}\sin(a+q) &= \cos q \sin a + \sin q \cos a, \\ \cos(a+q) &= \cos q \cos a - \sin q \sin a.\end{aligned} \quad (6)$$

Hence, if we can write $x = a_0 + q_1 + q_2 + \cdots + q_n$ for some set of angles a_0 and q_i, we can apply (6) repeatedly and eventually arrive at a formal representation for $\sin x$ and $\cos x$. Define

$$a_i = a_0 + q_1 + q_2 + \cdots + q_i = a_{i-1} + q_i.$$

Note that $x = a_n$. Assuming that $\sin q_i$ and $\cos q_i$ are known for each i, we have the procedure of **Table 3**.

Table 3.
Details behind the CORDIC algorithm for $\tan x$. The ellipses represent the computational application of the trigonometric identities (6).

i	$\sin a_i$	$\cos a_i$
0	$\sin a_0$	$\cos a_0$
1	$\sin(a_0 + q_1) = \ldots$	$\cos(a_0 + q_1) = \ldots$
2	$\sin(a_1 + q_2) = \ldots$	$\cos(a_1 + q_2) = \ldots$
$\vdots$	$\vdots$	$\vdots$
n	$\sin x = \sin(a_{n-1} + q_n) = \ldots$	$\cos x = \cos(a_{n-1} + q_n) = \ldots$

Whether the sequences in **Table 3** have any practical computational value depends on whether the repeated applications of the trigonometric identities in (6) can be done efficiently. That is, the computation of the intermediate sines and cosines must be *much* simpler than the direct computation of the sine and cosine. The fact that this can be done, as we will now see in detail, is what makes the CORDIC method practical.

Consider the general formulas from **Table 3**:

$$\begin{aligned}\sin(a_{i-1} + q_i) &= \cos q_i \sin a_{i-1} + \sin q_i \cos a_{i-1}, \\ \cos(a_{i-1} + q_i) &= \cos q_i \cos a_{i-1} - \sin q_i \sin a_{i-1}.\end{aligned} \quad (7)$$

Assume that $\sin a_{i-1}$ and $\cos a_{i-1}$ are known from the previous step in the iteration. Then the simplicity of the recurrence relations depends predominantly on the simplicity of $\sin q_i$ and $\cos q_i$, both in themselves and as multiplication factors in the current iteration. The heart of the method is to choose the q_i to obtain this simplicity.

When computations are done in decimal, multiplication by an integral power of 10 is just a shift of the decimal point in the other factor. An attempt to find an angle q such that both $\sin q$ and $\cos q$ are integral powers of 10 is fruitless. But finding an angle q for which the ratio of $\sin q$ to $\cos q$ is an integral power of 10 is straightforward.

Let $T_i = 10^{p_i}$ represent an integral (usually nonpositive) power of 10. Define

$$q_i = \arctan T_i, \qquad 0 < q_i < \pi/2.$$

From the fundamental identities illustrated by **Figure 3**, we note that if $w_i = \sqrt{1+T_i^2}$, then

$$\sin q_i = \frac{T_i}{w_i}, \qquad \cos q_i = \frac{1}{w_i}. \tag{8}$$

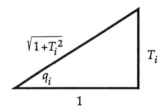

Figure 3. Triangle for fundamental identities.

Table 4 illustrates some possible choices of T_i. Notice that angles with different subscripts (2 and 3) need not be different in value.

Table 4.
Possible choices for T_i.

i	T	$q_i = \arctan T_i$	q_i	w_i	$\cos q_i$	$\sin q_i$
1	10^0	arctan 1	0.78540	1.41421	$1/w_1$	$1/w_1$
2	10^{-1}	arctan 0.1	0.09967	1.00499	$1/w_2$	$0.1/w_2$
3	10^{-1}	arctan 0.1	0.09967	1.00499	$1/w_3$	$0.1/w_3$
4	10^{-2}	arctan 0.01	0.01	1.00005	$1/w_4$	$0.01/w_4$

Consider again the recurrence relations of (7), but now in the particular case that $q_i = \arctan T_i$ and angle a_0 is so small that $\sin a_0 \approx a_0$ and $\cos a_0 \approx 1$.

Let
$$S_i = \sin a_i = \sin(a_0 + q_1 + \cdots + q_i),$$
$$C_i = \cos a_i = \cos(a_0 + q_1 + \cdots + q_i).$$

Then, using **(8)** extensively, we have the results in **Table 5**.

Table 5.
Steps in the CORDIC technique for $\tan x$.

$i=0$	S_0	$= \sin a_0 = a_0$	C_0	$= \cos a_0 = 1$
$i=1$	S_1	$= \sin(a_0 + q_1)$	C_1	$= \cos(a_0 + q_1)$
		$= \cos q_1 \sin a_0 + \sin q_1 \cos a_0$		$= \cos q_0 \cos a_0 - \sin q_1 \sin a_0$
		$= \frac{1}{w_1} \sin a_0 + \frac{T_1}{w_1} \cos a_0$		$= \frac{1}{w_1} \cos a_0 - \frac{T_1}{w_1} \sin a_0$
		$= \frac{1}{w_1}(S_0 + T_1 C_0)$		$= \frac{1}{w_1}(C_0 - T_1 S_0)$

The steps shown in detail for $i = 1$ may be paralleled for $i > 1$ to obtain the general recursion relations

$$S_i = \frac{1}{w_i}(S_{i-1} + T_i C_{i-1}), \qquad C_i = \frac{1}{w_i}(C_{i-1} - T_i S_{i-1}). \qquad (9)$$

These recurrence relations are formally simple but still not sufficiently efficient computationally to be useful. For although the multiplication by T_i can be done by shifting the decimal point, the division by w_i remains prohibitively costly computationally.

A slight modification of **(9)** will eliminate the problem of the divisions by w_i, but at the expense of changing the meanings of the recursive variables. The new recursion formulas are motivated by rewriting the elements of **(9)** with the w_i's as factors on the left, rather than as divisors on the right. That is,

$$\begin{aligned} S_0 &= a_0, & C_0 &= 1 \\ w_1 S_1 &= S_0 + T_1 C_0, & w_1 C_1 &= C_0 - T_1 S_0 \\ &= a_0 + T_1 & &= 1 - T_1 a_0 \\ &\vdots & &\vdots \end{aligned}$$

Now define the left sides of the relations above as the new recursive variables and generalize them as follows:

$$\begin{aligned} P_0 &= S_0 = a_0, & Q_0 &= C_0 = 1 \\ P_1 &= w_1 S_1, & Q_1 &= w_1 C_1. \end{aligned}$$

Letting $W_i = w_1 w_2 \cdots w_i$, for $i = 1, 2, \ldots, n$, put

$$P_i = W_i S_i, \qquad Q_i = W_i C_i, \qquad (10)$$

where the definitions of S_i, C_i, and a_i are still as given before. We now collect the last few remarks into the following:

Theorem. P_i and Q_i as defined in **(10)** satisfy the recursion relations

$$P_0 = a_0 \ (\approx \sin a_0), \quad Q_0 = 1 \ (\approx \cos a_0),$$
$$\text{and for } i > 1, \quad P_i = P_{i-1} + T_i Q_{i-1}, \quad Q_i = Q_{i-1} - T_i P_{i-1}. \tag{11}$$

Proof. Since $S_0 = a_0$, $P_0 = a_0$. Likewise, since $C_0 = 1$, $Q_0 = 1$. Below we do the proof for general P, by induction. The proof for Q is left as **Exercise 9**.

i) Proof for $i = 1$ (recall **(8)**):

$$\begin{aligned} P_1 &= w_1 S_1 \\ &= w_1 \sin(a_0 + q_1) \\ &= w_1 (\cos q_1 \sin a_0 + \sin q_1 \cos a_0) \\ &= w_1 \left(\frac{1}{w_1} \sin a_0 + \frac{T_1}{w_1} \cos a_0 \right) \\ &= P_0 + T_1 Q_0. \end{aligned}$$

ii) Proof for $i+1$, assuming true for i:

$$\begin{aligned} P_{i+1} &= W_{i+1} S_{i+1} \\ &= w_{i+1} W_i \sin(a_i + q_{i+1}) \\ &= w_{i+1} W_i (\cos q_{i+1} \sin a_i + \sin q_{i+1} \cos a_i) \\ &= w_{i+1} W_i \left(\frac{1}{w_{i+1}} \sin a_i + \frac{T_{i+1}}{w_{i+1}} \cos a_i \right) \\ &= W_i S_i + W_i T_{i+1} C_i \\ &= P_i + T_{i+1} Q_i. \end{aligned} \qquad \square$$

The application of the theorem is a CORDIC method for computing $\tan x$. Computationally, all that is required is multiplication performed by shifting, addition, and subtraction. Recall that

$$x = a_0 + q_1 + q_2 + \cdots + q_n = a_n,$$

so $P_n = W_n S_n = W_n \sin x$ and $Q_n = W_n C_n = W_n \cos x$. The algorithm is completed by dividing: $\tan x = P_n/Q_n$, independent of W_n.

Exercise

9. Complete the induction proof of the theorem above.

Computer and Calculator Computation of Elementary Functions

5.1.2 The CORDIC computation of sines and cosines

The keys to CORDIC computation of sines and cosines are the trigonometric identities

$$\sin x = \frac{\tan x}{\sqrt{1 + \tan^2 x}}, \qquad \cos x = \frac{1}{\sqrt{1 + \tan^2 x}}.$$

Since $\tan x = P_n/Q_n$, it follows that we may use the fast square-root routine to calculate sines and cosines:

$$\sin x = \frac{\tan x}{\sqrt{1 + \tan^2 x}} = \frac{P_n}{\sqrt{P_n^2 + Q_n^2}}, \qquad \cos x = \frac{1}{\sqrt{1 + \tan^2 x}} = \frac{Q_n}{\sqrt{P_n^2 + Q_n^2}}. \quad (12)$$

Example. Compute $\sin 0.99474$ and $\cos 0.99474$.

In the previous example, we obtained $\tan 0.99474$ as $P_4/Q_4 = 1.1979/0.7781$. We apply **(12)** to obtain $\sin 0.99474 = 0.83861$ and $\cos 0.99474 = 0.54473$.

Exercise

10. Compute $\sin 2$ and $\cos 2$ using the CORDIC algorithm. Let $a_0 = 0.00053$.

5.2 A CORDIC Algorithm for Arctangent

With a change in initialization and a clever reinterpretation, the CORDIC algorithm for solving

$$\text{find } t = \tan x, \text{ given } x \text{ (by finding } W \sin x \text{ and } W \cos x\text{)}$$

can be used to solve the inverse problem

$$\text{find } x = \arctan t, \text{ given } t.$$

We outline the CORDIC procedure for computing arctangent in **Algorithm 2** and illustrate it with an easy example. Then, in section 5.2.1, we derive the algorithm.

Algorithm 2.
The CORDIC algorithm for $\arctan t$. The T_i's are chosen so that $P_i \downarrow 0$.

$$\arctan t = a_n$$

where $a_0 = 0$
  $a_i = a_{i-1} + \arctan T_i$ $T_i \in \{1,\ 0.1,\ 0.01,\ \ldots\}$

  $P_0 = t$ $Q_0 = 1$
  $P_i = P_{i-1} - T_i Q_{i-1}$ $Q_i = Q_{i-1} + T_i P_{i-1}$

Example. Compute $\arctan 0.21349$.

Recall that

$$\begin{aligned}\arctan 1 &= 0.78540, \\ \arctan 0.1 &= 0.09967, \\ \arctan 10^{-j} &\approx 10^{-j} \text{ for } j \geq 2.\end{aligned}$$

We generate a sequence a_i by successively adding to a_{i-1} the largest angle $q_i (= \arctan T_i)$ that keeps P_i nonnegative. (We do not show explicitly the test for choice of q_i.) The process terminates when P_n is sufficiently close to zero. At that point $x = a_n$. Hence $x = \arctan 0.21349 = a_4 = 0.21034$. We show the successive calculations in **Table 6**.

Table 6.
The CORDIC algorithm for $\arctan 0.21349$.

	$P_0 = +0.21350$	$Q_0 = +1.00000$	$a_0 = 0$
$T_1 = 0.1$	$-T_1 Q_0 = -0.10000$ $P_1 = +0.11350$	$+T_1 P_0 = +0.02135$ $Q_1 = +1.02135$	$q_1 = 0.09967$ $a_1 = 0.09967$
$T_2 = 0.1$	$-T_2 Q_1 = -0.10214$ $P_2 = +0.01136$	$+T_2 P_1 = +0.01135$ $Q_2 = +1.03270$	$q_2 = 0.09967$ $a_2 = 0.19934$
$T_3 = 0.01$	$-T_3 Q_2 = -0.01033$ $P_3 = +0.00103$	$+T_3 P_2 = +0.00011$ $Q_3 = +1.03281$	$q_3 = 0.01000$ $a_3 = 0.20934$
$T_4 = 0.001$	$-T_4 Q_3 = 0.00103$ $P_4 = +0.00000$	$+T_4 P_3 = +0.00001$ $Q_4 = +1.03282$	$q_4 = 0.00100$ $a_4 = 0.21034$

5.2.1 Derivation of the CORDIC algorithm for arctangent

Let's first review **Algorithm 1**, the CORDIC algorithm for computing $t = \tan x$, with a view toward solving the inverse problem $x = \arctan t$. In that algorithm, we construct the sequence of angles a_i and the associated sequences P_i and Q_i, which are pairwise proportional to $\sin a_i$ and $\cos a_i$. Hence $\tan x = P_n/Q_n$. A fundamental aspect of the algorithm is the termination of the process. This occurs when $a_n = x$, which can be observed directly.

Any attempt to apply this algorithm to solve the inverse problem $x = \arctan t$ would require a different termination test. Even though the sequence of a's still converges to x, testing whether $a_n = x$ directly is not possible, since x, being the answer, is not known numerically. We might consider terminating the algorithm when $P_n/Q_n = t$, hence $a_n = x$. The trouble with this is not theoretical but computational; the divisions at each step violate the spirit of CORDIC, namely, to divide only by shifting.

A slight modification of the meaning of P_i and Q_i provides an algorithm for $\arctan t$. We introduce another sequence of angles, $\theta_0, \theta_1, \ldots, \theta_n$ (which are never seen in the calculations). The key is to choose the θ's so that testing whether $a_n = x$ is equivalent to testing whether $\tan \theta_n = 0$. We also introduce P_i and Q_i to be pairwise proportional to $\sin \theta_i$ and $\cos \theta_i$. In fact, the proportionality constants will be such that $Q_i \geq 1$, hence $P_i \geq \sin \theta_i$.

The construction is as follows:

$$\begin{aligned} a_0 + \theta_0 &= 0 + \theta_0 = x \\ (a_0 + q_1) + (\theta_0 - q_1) &= a_1 + \theta_1 = x \\ &\vdots \qquad\qquad\qquad \vdots \\ (a_{n-1} + q_n) + (\theta_{n-1} - q_n) &= a_n + \theta_n = x. \end{aligned}$$

Note that $\theta_0 = x$. This value is not known numerically, but what is known is $\tan \theta_0$, that is, t. At the point where $\theta_n = 0$, a_n will equal x, the required answer. Note further that $\tan \theta_n$ is zero.

In the initial step, we choose $P_0 = t$ and $Q_0 = 1$. An identity yields recurrence formulas for P_i and Q_i. Recall that $P_i/Q_i = \tan \theta_i$. The q_i are still the special angles $\arctan T_i$, chosen to drive the θ's toward zero. Now

$$\tan \theta_{i+1} = \tan(\theta_i - q_{i+1}) = \frac{\tan \theta_i - \tan q_{i+1}}{1 + \tan \theta_i \tan q_{i+1}} = \frac{P_i/Q_i - T_{i+1}}{1 + (P_i/Q_i)T_{i+1}} = \frac{P_i - Q_i T_{i+1}}{Q_i + P_i T_{i+1}}.$$

This shows that we should define:

$$P_{i+1} = P_i - Q_i T_{i+1}, \qquad Q_{i+1} = Q_i + P_i T_{i+1}.$$

(This definition assures that $Q_i \geq 1$.) Furthermore, we may dispense with an explicit concern for the auxiliary sequence of θ's. Since $\lim_{\theta \to 0}(\tan \theta)/\theta = \lim_{\theta \to 0}(\sin \theta)/\theta = 1$, then $\tan \theta$, $\sin \theta$, and θ all have essentially the same value near zero. Testing whether θ is sufficiently small is equivalent to testing whether $\sin\theta$ is small. Since $P_n \geq \sin\theta$, then P_n may be tested computationally.

Collecting these ideas yields **Algorithm 2**. In practice, it is not necessary to drive the P's completely to 0. We may make use of the fact that when x is small, $\tan x \approx x$. Suppose that P_i, Q_i, and a_i have been computed and that P_i is small. At this point, let $n = i+1$. Since $Q_{n-1} \geq 1$, then $\arctan(P_{n-1}/Q_{n-1}) \approx P_{n-1}/Q_{n-1}$. We may terminate the process by taking the last angle to be $q_n = P_{n-1}/Q_{n-1}$. With this choice, we have $P_n = 0$ and $a_n = a_{n-1} + P_{n-1}/Q_{n-1}$.

Exercise

11. Compute $\arctan 2$ using CORDIC.

Computer and Calculator Computation of Elementary Functions   29

6. Solutions to the Exercises

1. The truncated series of degree n for e^x is $\sum_{k=0}^{n} x^k/k!$. By Taylor's theorem, this has remainder $R_n(c) = e^c x^{n+1}/(n+1)!$, where c depends upon x and lies between 0 and x.

   a) Since $e^1 < 2.8$, to obtain six-digit accuracy, $|R_n(c)|$ must be bounded by 0.5×10^{-6}. One can verify that $n = 10$.

   b) Since $e^{10} < 23,000$, to obtain six-digit accuracy, $|R_n(c)|$ must be bounded by 0.05. One can verify that $e^{10} 10^{36}/36! \approx 0.06$ and $e^{10} 10^{37}/37! \approx 0.017$. Hence $n = 36$.

2. Use a function grapher to see that the maximum occurs at $x = 1$.

3. Write the sum as $0.23 \times 10^5 + 0.00000000053 \times 10^5 = 0.23000000053 \times 10^5$. Note, however, that addition on a machine with eight-digit floating-point precision will produce instead the result 0.23000000×10^5.

4. $a_0 = 0$, $R_0 = 27.17954$. For $j = 0$, we have

$$\begin{aligned} 10^{-1} \times 5R_1 &= 10^{-1} \times 5R_0 - \sum_{i=1}^{b}\left(0 + (i-1)|5 \times 10^{-1}\right) \\ &= \frac{1}{2}(27.17954) - \sum_{i=1}^{b}\left((i-1)|5 \times 10^{-1}\right) \\ &= 13.58977 - (0.5 + 1.5 + 2.5 + 3.5 + 4.5) \\ &= 1.08977. \end{aligned}$$

Hence $b = 5$ (that is, five subtractions) and $a_1 = 5$. With $j = 1$, we have

$$\begin{aligned} 5R_2 &= 5R_1 - \sum_{i=1}^{b}\left(5 + (i-1)|5 \times 10^{-2}\right) \\ &= 10.8977 - (5.05 + 5.15) \\ &= 0.6977. \end{aligned}$$

Hence $b = 2$ and $a_2 = 5.2$. The rest continues as in the **Example**.

5. a) $a_0 = 0.54976933$, $a_1 = 0.5477263675$, $a_2 = 0.5477225575$ (only two iterations).

   b) $a_0 = 0.8319999$, $a_1 = 0.8366730775$, $a_2 = 0.8366600265$ (only two iterations).

6. $p(x) = 1 + x \ln 2 + x^2 (\ln 2)^2/2 + x^3 (\ln 2)^3/6 + x^4 (\ln 2)^4/24$, with an upper bound for the error on the interval $\left[\frac{-1}{16}, \frac{1}{16}\right]$ given by $(\ln 2)^5 2^{1/16} (1/16)^5/60 = 2.655 \times 10^{-9} < 3 \times 10^{-9}$.

25

7. a) The Maclaurin series for $\ln(1+x)$ is $x - \frac{1}{2}x^2 + \frac{1}{3}x^3 - \frac{1}{4}x^4 + \cdots$, valid for $|x| < 1$. The series for $\ln(1-x)$ is $-\left(x + \frac{1}{2}x^2 + \frac{1}{3}x^3 + \frac{1}{4}x^4 + \cdots\right)$. Therefore, for $|x| < 1$,

$$\ln\left(\frac{1+x}{1-x}\right) = \ln(1+x) - \ln(1-x) = 2\left(x + \frac{x^3}{3} + \frac{x^5}{5} + \cdots\right).$$

b) $\dfrac{1+V}{1-V} = \dfrac{1+\left(\frac{X-A}{X+A}\right)}{1-\left(\frac{X-A}{X+A}\right)} = \dfrac{2X}{2A} = \dfrac{X}{A}.$

Therefore, for $|V| < 1$,

$$\ln\left(\frac{1+V}{1-V}\right) = \ln\left(\frac{X}{A}\right) = \ln X - \ln A = 2\left(V + \frac{V^3}{3} + \frac{V^5}{5} + \cdots\right).$$

8. a) From (3), $\ln 1.001 = 0.001 - (0.001)^2/2 + (0.001)^3/3 = 0.0009995.$
   b) Let $V = (0.78 - 0.75)/(0.78 + 0.75) = 0.019608$. From (5),

$$\ln 0.78 = -0.28768 + 2(0.019608)\left(1 + \frac{(0.019608)^2}{3} + \frac{(0.019608)^4}{5}\right)$$
$$= -0.248459.$$

9. i) For $i = 1$:
$$\begin{aligned}Q_1 &= w_1 C_1 \\ &= w_1 \cos(a_0 + q_1) \\ &= w_1(\cos a_0 \cos q_1 - \sin a_0 \sin q_1) \\ &= w_1\left((\cos a_0)\frac{1}{w_1} - (\sin a_0)\frac{T_1}{w_1}\right) \\ &= Q_0 - T_1 P_0.\end{aligned}$$

ii) Proof for $i+1$, assuming true for i:
$$\begin{aligned}Q_{i+1} &= w_{i+1} C_{i+1} \\ &= w_{i+1} W_i \cos(a_i + q_{i+1}) \\ &= w_{i+1} W_i(\cos a_i \cos q_{i+1} - \sin a_i \sin q_{i+1}) \\ &= w_{i+1} W_i\left((\cos a_i)\frac{1}{w_{i+1}} - (\sin a_i)\frac{T_{i+1}}{w_{i+1}}\right) \\ &= Q_i - T_{i+1} P_i.\end{aligned}$$

10. $2 = 2\arctan 1 + 4\arctan 0.1 + 3\arctan 0.01 + 0.00053$. The intermediate computations are shown in **Table 7**. Using the identities **(12)**, we obtain $\sin 2 = 0.909297$ and $\cos 2 = -0.41615$.

11. After the last line of computations in **Table 8**, we claim that P_6 is small. Hence $\arctan 2 = a_7 = a_6 + P_6/Q_6 = 1.10715$.

Table 7.
Computations for **Exercise 10**.

i	T_i	P_i	Q_i
0		0.00053	+1.00000
1	1	1.00053	+0.99947
2	1	2.00000	−0.00106
3	0.1	1.99989	−0.20106
4	0.1	1.97978	−0.40105
5	0.1	1.93968	−0.59990
6	0.1	1.87969	−0.79300
7	0.01	1.87176	−0.81180
8	0.01	0.86364	−0.83051
9	0.01	0.85534	−0.84915

Table 8.
Computations for **Exercise 11**.

i	T_i	P_i	Q_i		a_i
0		2.00000	1.00000		0
1	1	1.00000	3.00000	$\arctan 1 =$	0.78540
2	0.1	0.70000	3.10000	$a_1 + \arctan 0.1 =$	0.88507
3	0.1	0.39000	3.17000	$a_2 + \arctan 0.1 =$	0.98474
4	0.1	0.07300	3.20900	$a_3 + \arctan 0.1 =$	1.08441
5	0.01	0.04091	3.20973		1.09441
6	0.01	0.00881	3.21014		1.10441

References

There are many aspects to computer function approximation that we could not discuss in a Module of this size. The brief comments that we provide below with the references could serve as a starting point toward further investigation. Mention of the CORDIC algorithm in numerical analysis or numerical methods texts is rare. Hence there are few such entries in this bibliography.

We call the reader's attention to two particular areas of CORDIC. First, there is an enlightening geometric interpretation to the CORDIC sequences. Second, CORDIC algorithms for all of the elementary functions have essentially the same form. The main differences among them are initialization and the particular table of values that are stored. Hence, it is possible to construct a *unified algorithm* for CORDIC computation.

Abramowitz, M., and I.A. Stegun, eds. 1965. *Handbook of Mathematical Functions*. New York: Dover.

A standard reference book for special functions.

Cody, W J., and W. Waite. 1980. *Software Manual for the Elementary Functions*. Englewood Cliffs, NJ: Prentice-Hall.

Provides accuracy tests for the elementary functions.

Duncan, Ray. 1989–1990. Arithmetic routines for your computer programs, Parts 1–7. *PC Magazine* 8(19) (14 November 1989): 423–429; 8(20) (28 November 1989): 345–352; 8(21) (12 December 1989): 337–354; 8(22) (26 December 1989): 271–277; 9(1) (16 January 1990): 311–332; 9(3) (13 February 1990): 297–307; 9(5) (13 March 1990): 353–370.

These columns describe the IEEE standards for floating-point representation, integer and floating-point arithmetic routines, and computation of elementary functions, using math coprocessors.

Egbert, W.E. 1977–1978. Personal calculator algorithms I: Square roots. II. Trigonometric functions. III. Inverse trigonometric functions. IV. Logarithmic functions. *Hewlett-Packard Journal* 28(9) (May 1977): 22–24; 28(10) (June 1977): 17–20; 29(3) (November 1977): 22–23; 29(8) (April 1978): 29–32.

These columns describe in detail the CORDIC algorithms as developed for the HP-35.

Fike, C.T. 1968. *Computer Evaluation of Elementary Functions*. Englewood Cliffs, NJ: Prentice-Hall.

A classic.

Gerald, C.F., and P.O. Wheatley. 1989. *Applied Numerical Analysis.* 4th ed. Reading, MA: Addison-Wesley.

Discussion of Chebyshev and Padé approximations, mention of CORDIC.

Glass, L. Brent. 1990. Math coprocessors. *Byte* 15(1) (January 1990): 340.

Mentions a math coprocessor recently introduced by Cyrix which uses polynomial approximations rather than CORDIC.

Hart, John F. 1978. *Computer Approximations.* Huntington, NY: Krieger.

Source of analytic approximations.

Kropa, J.C. 1978. Calculator algorithms. *Mathematics Magazine* 51(2): 106–109.

Includes geometric interpretation of CORDIC.

Ruckdeschel, F.R. 1981. *Basic Scientific Subroutines.* Vol. 2. New York: Byte/McGraw-Hill.

Contains a discussion of CORDIC and gives source listings of BASIC programs that imitate CORDIC.

Schelin, C.W. 1983. Calculator function approximation. *American Mathematical Monthly* 90(5): 317–325.

Discussion of unified CORDIC algorithm, with an extensive bibliography.

Schmid, Herman. 1974. *Decimal Computations.* New York: Wiley.

Discussion of CORDIC and other techniques, with emphasis on hardware.

Spafford, E., and J. Flaspohler. 1985. A report on the accuracy of some floating point math functions on selected computers. Technical Report GIT–ICS 85/06. Atlanta, GA: School of Information and Computer Science, Georgia Institute of Technology.

Comparison of accuracy in library routines for a large class of computers.

Swartzlander, E.E., ed. 1980. *Computer Arithmetic.* Stroudsburg, PA: Dowden, Hutchinson & Ross.

A collection of papers, including those of Volder and Walther.

VAX Programming, Vol. 5A: Run-Time Library. 1988. Maynard, MA: Digital Equipment Corporation.

Volder, J.E. 1959. The CORDIC trigonometric computing technique. *IRE Transactions on Electronic Computation* EC–8: 330–334 (September 1959).

The paper that started it all (if one doesn't count Briggs).

Walther, J.S. 1971. A unified algorithm for elementary functions. *Spring Joint Computer Conference Proceedings* (1971): 379–385.

A unified CORDIC algorithm with hardware emphasis.

Acknowledgments

The CORDIC section of this Module is the result of assistance from many sources. Special thanks to Ron Tabor of Hewlett-Packard and Jeff Crumb and C.B. Wilson of Texas Instruments, who responded to our inquiries and furnished us with information about their machines. The authors also wish to thank the referees and the editor for many helpful suggestions.

About the Authors

Jim Delaney is a professor in the Dept. of Mathematics and Computer Science at Xavier University, where he has been a member since 1963. Prior to that, he was employed for six years as a programmer-analyst and applied mathematician by the General Electric Co. in Evendale, Ohio. He received a Ph.D. in mathematics from the University of Cincinnati. His interests include computer science education, computer accuracy, and applied mathematics.

Dick Pulskamp is an assistant professor of mathematics at Xavier. He received his Ph.D. in Mathematics in 1988 from the University of Cincinnati. While his main area of interest is mathematical statistics, he is also interested in problems of numerical computations.

UMAP

Modules in Undergraduate Mathematics and its Applications

Published in cooperation with the Society for Industrial and Applied Mathematics, the Mathematical Association of America, the National Council of Teachers of Mathematics, the American Mathematical Association of Two-Year Colleges, The Institute of Management Sciences, and the American Statistical Association.

Module 712

The Drag Force on a Sphere

H. Edward Donley

Applications of Mechanics and Fluid Dynamics

COMAP, Inc., Suite 210, 57 Bedford Street, Lexington, MA 02173 (617) 862–7878

INTERMODULAR DESCRIPTION SHEET:	UMAP Unit 712
TITLE:	The Drag Force on a Sphere
AUTHOR:	H. Edward Donley Dept. of Mathematics Indiana University of Pennsylvania Indiana, PA 15705
MATHEMATICAL FIELD:	Differential Equations, Mathematical Modeling
APPLICATION FIELD:	Mechanics, Fluid Dynamics
TARGET AUDIENCE:	Students who have studied calculus through simple differential equations.
ABSTRACT:	This Module analyzes the drag force on a sphere moving through a fluid, by applying dimensional analysis to reduce the number of variables, experimental results to find a relationship between the drag coefficient and the Reynolds number, and the resulting log-log graph to develop two models for the drag force. These models are then used to derive differential equations for spheres falling through fluids.
PREREQUISITES:	Calculus through simple differential equations, including familiarity with separation of variables, logarithms and exponentials, and hyperbolic trigonometric functions.

RELATED UNITS:	Unit 526: Dimensional Analysis,
	    by Frank Giordano and Maurice Weir.
	
	Unit 564: Keeping Dimensions Straight,
	    by George E. Strecker.

©Copyright 1991, 1992 by COMAP, Inc. All rights reserved.

COMAP, Inc., Suite 210, 57 Bedford Street, Lexington, MA 02173
(800) 77–COMAP = (800) 772–6627,   (617) 862–7878

The Drag Force on a Sphere

H. Edward Donley
Dept. of Mathematics
Indiana University of Pennsylvania
Indiana, PA 15705

Table of Contents

1. INTRODUCTION 1

2. REDUCTION OF THE NUMBER OF VARIABLES 3
   2.1 The Need for Reducing the Number of Variables 3
   2.2 The Drag Coefficient and the Reynolds Number 5

3. GRAPH OF DRAG COEFFICIENT VS. REYNOLDS NUMBER 7
   3.1 Log-Log Graphs 7
   3.2 The Graph of C_D vs. R 9

4. TWO MODELS FOR THE DRAG FORCE 11

5. THE MOTION OF A SPHERE THROUGH A FLUID 11
   5.1 Development of the Differential Equations 11
   5.2 Solutions of the Differential Equations 12
   5.3 Comparison of the Two Models 14
   5.4 An Example: Sand Settling in Water 14

6. CONCLUSION 17

7. APPENDIX I: TABLE OF PHYSICAL CONSTANTS 17

8. APPENDIX II: DIMENSIONAL ANALYSIS 18

9. SAMPLE EXAM 20

10. SOLUTIONS TO THE EXERCISES 21

11. ANSWERS TO THE SAMPLE EXAM 31

    REFERENCES 32

    ABOUT THE AUTHOR 32

MODULES AND MONOGRAPHS IN UNDERGRADUATE MATHEMATICS AND ITS APPLICATIONS (UMAP) PROJECT

The goal of UMAP is to develop, through a community of users and developers, a system of instructional modules in undergraduate mathematics and its applications, to supplement existing courses and from which complete courses may be built.

The Project was initially funded by a grant from the National Science Foundation and has been guided by a National Advisory Board of mathematicians, scientists, and educators. UMAP is now supported by the Consortium for Mathematics and Its Applications (COMAP), Inc., a non-profit corporation engaged in research and development in mathematics education.

COMAP Staff

Paul J. Campbell	Editor
Solomon Garfunkel	Executive Director, COMAP
Laurie W. Aragón	Development Director
Philip A. McGaw	Production Manager
Roland Cheyney	Project Manager
Laurie M. Holbrook	Copy Editor
Dale Horn	Design Assistant
Rob Altomonte	Distribution Coordinator
Sharon McNulty	Executive Assistant

1. Introduction

Fluids, whether liquids or gases, resist the motion of objects traveling through them. Every time you go swimming or put your hand out of the window of a moving car, you work against this resistance, called a *drag force*. What causes this drag force and how do we construct a mathematical model for it?

We will answer these questions for the case of a sphere moving through a fluid, or equivalently, a fluid flowing past a sphere.

The drag force is in the opposite direction of the velocity of the object, tending to retard its motion. The drag force has several sources, as we will see in Section 2. The parameters affecting this force are:

- the velocity of the object (or of the fluid),
- the size of the object,
- the shape of the object,
- the density of the fluid, and
- the viscosity of the fluid.

Viscosity is the internal frictional force per unit area of one layer of fluid passing over another layer. Viscosity is caused by interactions of particles of fluid at the molecular level. Imagine one small bit of fluid, called a *fluid element*, passing over another fluid element. While the two elements are in contact, some of the faster molecules from the first element will interchange with slower molecules from the second element. This causes the faster element to slow down and the slower element to speed up, as you would expect from a frictional force. When you stir a pitcher of lemonade, it is viscosity that causes the lemonade to come to rest eventually after you withdraw the spoon, because the lemonade adhering to the sides of the pitcher slows down the moving lemonade in the interior of the pitcher. Appendix I lists the densities and viscosities of the fluids referred to in this unit.

We will avoid discussing the complicated effect of the shape of an object on the drag force:

(A). *We restrict our study to spherical objects.*

Studies analogous to this one can be performed for other shapes. Diamond [1989] compares the shapes of humans and cats and describes the effects their shapes have on their drag force in air. Humans do not have a good internal gyroscope in their inner ears, so they tend to fall head first, feet first, or tumbling haphazardly. In contrast, cats fall feet first and probably spread their legs when falling from great heights. This orientation gives them a relatively larger cross-sectional area, which increases the drag force. The record for a cat falling from a New York City building is held by Sabrina, who fell 32 stories onto a sidewalk and only suffered a broken tooth and mild chest injuries! (The data were collected at the Animal Medical Center in Manhattan and were not part of an intentional experiment. Diamond points out that the empirical method can be used even in situations in which controlled experiments are impossible or unethical.)

We will also restrict our attention to problems in which:

(B). *The density of the fluid is constant.*

(C). *The velocity is significantly less than the speed of sound in the fluid, thus eliminating the possibility of sonic shock waves forming.*

Physical problems that satisfy our assumptions can be used both to test the validity of our mathematical model and to guide us in developing the model. Some applications satisfying our assumptions are:

- a rubber ball dropped from a cliff,
- water droplets dripping from a rooftop,
- a helium balloon rising into the air,
- sediment settling in a pool of water, and
- a baseball flying through the air.

In 1906 R.A. Millikan determined the electrical charge of an electron by using small electrically-charged oil droplets moving vertically through air in an electric field [Millikan 1963]. He used the velocity of the droplets to find their diameters and masses, and hence their charges.

On the other hand, the following applications do *not* satisfy our assumptions:

- a lead ball shot from a musket, since it travels faster than the speed of sound;
- the flight of an airplane, since an airplane is not a sphere; and
- the flight of a comet through the earth's atmosphere, since it travels faster than the speed of sound and travels through air of varying density.

Changing the parameters can have a significant effect on the nature of the fluid flow past a sphere. **Figure 1** shows the flow field for water flowing to the right past a sphere of diameter 2 cm. At very low velocities (**Figure 1a**), the downstream flow lines are a mirror image of the upstream flow lines. At slightly higher velocities (**Figure 1b**), a wide wake forms behind the sphere. At high velocities (**Figure 1c**), a thin layer of turbulence forms around the sphere and the wake, now narrower, also contains some turbulence. The fluid within the wake becomes more turbulent as the velocity increases.

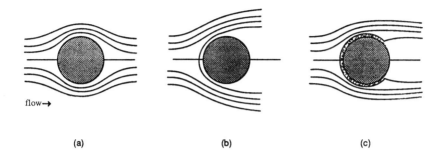

Figure 1. The flow of water past a sphere of diameter 2 cm can dramatically change character as the velocity changes. At very low velocities (a), there is no wake behind the sphere. At intermediate velocities (b), a boundary layer forms in front of the sphere and a wake forms behind it. At high velocities (c), the boundary layer becomes turbulent and the wake becomes narrower. The velocities for the three flows shown above are 2.5×10^{-3} cm/sec, 50 cm/sec, and 500 cm/sec, respectively.

In Section 2, we will describe how the parameters affect the drag force. The details of the dimensional analysis for Section 2 are in Appendix II; they may be skipped without loss of continuity. In Section 3, we present experimental data that are used in Section 4 to derive two models for the drag force. Then, in Section 5, we incorporate these models for the drag force into differential equations describing the vertical motion of a sphere subjected to gravity and drag. You will then be asked to solve the differential equations and compare the models.

2. Reduction of the Number of Variables

2.1 The Need for Reducing the Number of Variables

In Section 1, we mentioned that the drag force, F_D, depends on

- the velocity v of the sphere,
- the diameter D of the sphere,
- the density ρ of the fluid, and
- the viscosity μ of the fluid.

There is no known analytical method for finding F_D, except in special cases, so it must be determined experimentally.

Unfortunately, the drag force is a function of four variables, so it appears that many experiments would have to be performed to get a reasonably good description of it. For example, suppose we wanted to find F_D for 10 values of each variable. How many experiments would we have to conduct? We will answer this question by first examining simpler situations. We would need to conduct 10 experiments to model a function of one variable. Given a function of two variables, for each of the 10 values of the second variable, we would need to conduct experiments for 10 values of the first variable, for a total of 100 experiments. Given a function of three variables, for each of the 10 values of the third variable, we would need to conduct experiments for the 100 possible combinations of values of the first two variables, for a total of 10^3 experiments. Experimentally modeling the drag force, which is a function of four variables, would require 10^4 experiments! The powerful technique of dimensional analysis shows that these five variables can be combined into *two* variables without any loss of ability to describe the drag force. These two variables, derived in Appendix II, are the *drag coefficient* C_D and the *Reynolds number R*, defined by

$$C_D = \frac{F_D}{\frac{1}{2}\rho v^2 A} \quad \text{and} \quad R = \frac{\rho v D}{\mu}, \quad (1)$$

where A is the cross-sectional area of the sphere.

Both C_D and R are dimensionless. It can be shown that any model for the drag force can be expressed as a relationship between, and involving only, these two dimensionless parameters; that is,

$$\frac{F_D}{\frac{1}{2}\rho v^2 A} = f\left(\frac{\rho v D}{\mu}\right), \quad \text{or} \quad F_D = \frac{1}{2}\rho v^2 A f\left(\frac{\rho v D}{\mu}\right).$$

The function f must be determined experimentally. However, to do so requires far fewer experiments than to determine the original function of four variables.

Exercise

1. Determine what speed a sphere of diameter 5.00 cm would have to have, moving through glycerine, in order to have the same Reynolds number as a sphere of diameter 1.00×10^{-2} cm moving at 0.200 cm/sec through water at 20°C. See Appendix I for needed measurements of physical properties of fluids.

   Note: Since these two systems have the same Reynolds number, the flow pattern of one system is a scaled-down replica of the other system. An experimenter can determine the flow pattern for one system by constructing an analogous, more tractable system. High velocities, and very large

2.2 The Drag Coefficient and the Reynolds Number

The drag coefficient is the ratio of the drag force and the inertial force. The *inertial* (or *pressure*) *force* is the dynamic pressure times the cross-sectional area,

$$\text{inertial force} = \rho v^2 \times \frac{\pi}{4} D^2.$$

You can think of this as the force the moving sphere encounters in pushing fluid molecules out of its path.

The Reynolds number is the most commonly used of several standard dimensionless parameters in fluid mechanics. It is the ratio of the inertial force and the viscous force. The *viscous* (or *friction*) *force* is the viscous stress times the cross-sectional area,

$$\text{viscous force} = \frac{\mu v}{D} \times \frac{\pi}{4} D^2.$$

This force is due to the friction of the fluid molecules sliding alongside the sphere.

A plate moving through a fluid, with its orientation perpendicular to the flow, as in **Figure 2a**, would be subjected to a large pressure force resulting from the low pressure region behind the plate; but the friction force would be small. On the other hand, the flow past a plate oriented parallel to the flow, as in **Figure 2b**, would be dominated by the friction force. The flow past a sphere, having a curved surface, shares characteristics of both flows in **Figure 2**. In **Figure 3**, pressure forces dominate at point A; at points B and C, friction forces dominate.

The relative importance of each force for the entire sphere depends on the Reynolds number. At small Reynolds numbers, the friction force is much greater than the pressure force; and at large Reynolds numbers, the pressure force is much greater. Locomotion at low Reynolds numbers, where inertia is negligible compared to viscosity, differs radically from locomotion at high Reynolds numbers. This is why microscopic organisms have different shapes and locomotive apparatus from large aquatic and avian organisms (very small diameters implies low Reynolds numbers, provided that all other parameters are held constant) (see Purcell [1977]). The flows in **Figure 1** have Reynolds numbers 0.5, 10^4, and 10^6, respectively.

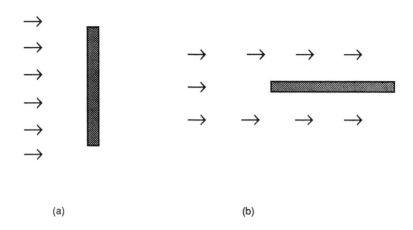

Figure 2. The plate in (a) is subjected to a large pressure force, while the plate in (b) is subjected to a large friction force.

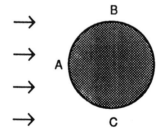

Figure 3. The pressure force is most important at point A, where the conditions are similar to those in Figure 2a. The friction force is most important at points B and C, which are similar to Figure 2b.

3. Graph of Drag Coefficient vs. Reynolds Number

In this section, we will present experimental results for the function f derived in our dimensional analysis of Section 2. The nature of the relationship between the drag coefficient and Reynolds number changes as these variables range over several orders of magnitude, so that it is useful to graph $\log C_D$ vs. $\log R$ instead of C_D vs. R. Later in this section, we will see another good reason for using logarithms in the graph. But first, we must learn something about logarithmic graphs.

3.1 Log-Log Graphs

Consider the data set of (x, y) values in **Table 1**. The graph of these

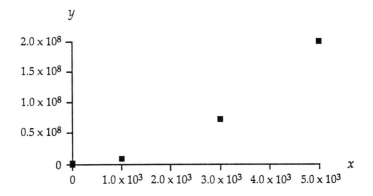

Figure 4. Because the data values vary greatly in magnitude, six of the data points coalesce.

ordered pairs is shown in **Figure 4**. Notice that the first four points nearly coalesce because our numbers vary greatly in magnitude. If we were to draw a curve to fit these ordered pairs approximately, we would probably get a good fit for the large values; but the relative error in the fit to the small values could be unacceptable. We would have difficulties in capturing the trends in both the large and the small values.

Since the logarithm of a large number is a much smaller number and the logarithm of a small positive number is a number that is much larger in magnitude, we can take the logarithm (we use base 10) of the values and get numbers that are all about the same order of magnitude. The logarithms of the data values are also given in **Table 1**, and the corresponding graph is shown in **Figure 5**.

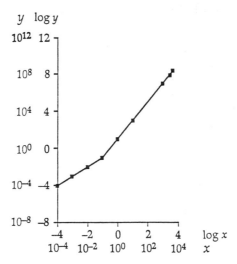

Figure 5. The log-log graph spreads the points more evenly.

The points no longer coalesce, and our graph now illustrates the behavior of the data over many orders of magnitude. This graph is called a *log-log graph*. We can read some important information from this graph. Notice that the graph appears to be piecewise-linear. For $x \geq 10^{-1}$, the points lie on a line with slope 2 and intercept 0.903. Thus, the equation relating x and y is

$$\begin{aligned} \log y &= 2\log x + 0.903 \\ &= \log x^2 + \log 10^{0.903} \\ &= \log(8.00x^2). \end{aligned}$$

Thus, $y = 8.00x^2$. The slope of the line in the log-log graph turns out to be the power of x in the equation for y. For $x \leq 10^{-1}$, the points lie on a line with slope 1 and intercept -0.097; you should show that this implies that $y = (8.00 \times 10^{-1})\,x$. If we had not transformed the data using logarithms, we would not have discovered that the data behave linearly for small x and quadratically for large x. The transformation also allowed us to find equations for the linear and quadratic parts of the function. Experimental data are often displayed in a log-log graph; then an empirical model can be constructed by fitting the points with a line or a piecewise linear function.

Exercise

2. W. Styś [1963] showed a correlation between the number of children in Polish peasant families and the size of their farms. The data for 8,505 families were grouped according to farm size (in hectares), and the average number of children was calculated for each farm size. The results

are shown in the log-log graph in **Figure 6**. Draw a line approximating the trend and find an equation for the number of children as a function of the farm size.

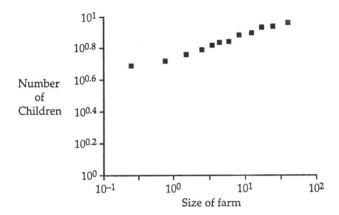

Figure 6. The relation between the size of a farm in Poland and the number of children in the family.

3.2 The Graph of C_D vs. R

The drag coefficient has been measured for spheres in fluid flows at various Reynolds numbers. The results are shown in the log-log graph in **Figure 7**.

The first section of the graph, for $R < 0.5$, corresponds to flows like that in **Figure 1a**. The theoretical drag coefficient for these low Reynolds number flows was derived by G.G. Stokes in 1851. He predicted that $C_D = 24/R$; in **Exercise 3**, you will be asked to compare this prediction with the experimental results.

As the Reynolds number increases, a wake develops behind the sphere. This wake is wide for $10^3 < R < 10^5$, as illustrated in **Figure 1b**. The large dip in C_D between $R = 10^5$ and $R = 10^6$ corresponds to the formation of a turbulent boundary layer in front of the sphere, accompanied by a narrower wake behind the sphere, as in **Figure 1c**. This narrower wake causes the drag force to actually decrease. Golf balls are corrugated in order to induce a turbulent boundary layer and reduce the drag force. The seams on a baseball allow a knuckleball to oscillate between having a turbulent and a non-turbulent boundary layer, thus causing erratic changes in speed and lift. Airplane wings have sharp trailing edges in order to reduce or eliminate the wake. The graph in **Figure 7** stops at $R = 10^6$ because it is difficult to obtain reliable experimental results for $R > 10^6$.

Table 1.
Artificial data covering many orders of magnitude.

x	y	$\log x$	$\log y$
10^{-4}	8×10^{-5}	-4	-4.097
10^{-3}	8×10^{-4}	-3	-3.097
10^{-2}	8×10^{-3}	-2	-2.097
10^{-1}	8×10^{-2}	-1	-1.097
1	8	0	0.903
10	800	1	2.903
10^3	8×10^6	3	6.903
3×10^3	7.2×10^7	3.477	7.857
5×10^3	2×10^8	3.699	8.301

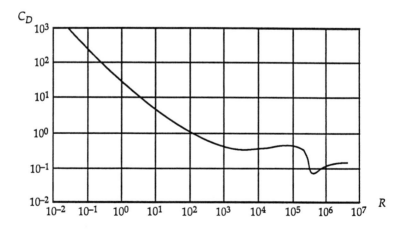

Figure 7. Experimentally determined relationship between the two dimensionless parameters, drag coefficient C_D and Reynolds number R.

Exercises

3. Draw a line segment to approximate the experimental data for $R < 0.5$. Find an equation for this line and solve the equation for C_D. How well do the experimental data match Stokes's theory?

4. Approximate C_D with a constant function for $10^3 < R < 10^5$.

4. Two Models for the Drag Force

Our two approximations for C_D in **Exercises 3** and **4** allow us to derive two formulas for the drag force F_D. From our original definition of the drag coeffiicent, we have

$$C_D = \frac{F_D}{\frac{1}{2}\rho v^2 A}, \quad \text{hence} \quad F_D = \frac{1}{2}\rho v^2 A C_D. \tag{2}$$

According to Stokes's theory, for $R < 0.5$, $C_D = 24/R$. Substituting this, we have

$$F_D = \frac{12\rho v^2 A}{R}.$$

But $R = \rho v D/\mu$, so

$$F_D = 12\rho v^2 A \times \frac{\mu}{\rho v D} = \frac{12\mu A v}{D} = \frac{12\mu \left(\frac{\pi}{4}D^2\right)v}{D} = 3\pi \mu D v$$
$$= k_1 v.$$

For $10^3 < R < 10^5$, C_D is approximately constant, so

$$F_D = \frac{1}{2}\rho v^2 A C_D = k_2 v^2.$$

Thus, depending on the Reynolds number, the drag force can be proportional either to v or to v^2.

5. The Motion of a Sphere through a Fluid

5.1 Development of the Differential Equations

Suppose that a sphere is moving vertically through a fluid under the influence of gravity and a drag force, with $R < 0.5$. Let $y(t)$ and $v(t)$ be the height and vertical velocity, respectively, of the sphere at time t. The forces acting on the sphere are shown in **Figure 8**. (Recall that the direction of the drag force is opposite to the direction of the motion.)

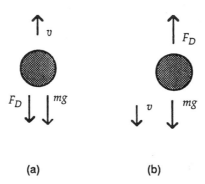

Figure 8. The gravitational and drag forces acting on a sphere moving (a) upward and (b) downward, through a fluid.

Newton's Second Law of Motion states that the mass times the acceleration is equal to the sum of the forces. So

$$\text{(for } R < 0.5\text{)} \qquad m\frac{dv}{dt} = -mg - k_1 v. \qquad (3)$$

The gravitational force on the right side of the equation is negative because its direction is opposite the direction of increasing y. There is a minus sign in front of the drag force because the drag force and v are in opposite directions.

If instead we have $10^3 < R < 10^5$, then $F_D = k_2 v^2$. So for an upwardly moving sphere,

$$\text{(for } 10^3 < R < 10^5\text{, sphere moving upward)} \qquad m\frac{dv}{dt} = -mg - k_2 v^2, \qquad (4)$$

and for a downwardly moving sphere,

$$\text{(for } 10^3 < R < 10^5\text{, sphere moving downward)} \qquad m\frac{dv}{dt} = -mg + k_2 v^2. \qquad (5)$$

5.2 Solutions of the Differential Equations

We represent the initial velocity of the sphere, at time $t = 0$, by v_0, and its location then by y_0.

For the linear drag force, with $R < 0.5$, the solution of the differential equation is

$$v = -\frac{mg}{k_1} + \left(v_0 + \frac{mg}{k_1}\right) e^{\frac{-k_1}{m} t}. \qquad (6)$$

Integrating with respect to t, we get

$$y = -\frac{mg}{k_1}t + \left(v_0 + \frac{mg}{k_1}\right)\left(e^{\frac{-k_1}{m}t} - 1\right) + y_0. \tag{7}$$

For an upwardly moving sphere under the influence of a quadratic drag force, with $10^3 < R < 10^5$, the velocity and displacement are

$$v = \sqrt{\frac{mg}{k_2}} \tan\left(-\sqrt{\frac{k_2 g}{m}}\, t + \arctan\left(\sqrt{\frac{k_2}{mg}}\, v_0\right)\right), \tag{8}$$

$$y = y_0 + \frac{m}{k_2} \ln\left|\sqrt{1 + \frac{k_2}{mg} v_0^2}\, \cos\left(\sqrt{\frac{k_2 g}{m}}\, t + \arctan\left(\sqrt{\frac{k_2}{mg}}\, v_0\right)\right)\right|. \tag{9}$$

For a falling sphere with a quadratic drag force, we have

$$v = \sqrt{\frac{mg}{k_2}} \tanh\left(-\sqrt{\frac{k_2 g}{m}}\, t + \arctan\left(\sqrt{\frac{k_2}{mg}}\, v_0\right)\right), \tag{10}$$

$$y = y_0 - \frac{m}{k_2} \ln\left|\sqrt{1 - \frac{k_2}{mg} v_0^2}\, \cosh\left(-\sqrt{\frac{k_2 g}{m}}\, t + \operatorname{arctanh}\left(\sqrt{\frac{k_2}{mg}}\, v_0\right)\right)\right|. \tag{11}$$

The velocities of objects falling through fluids do not increase without bound; they approach a limiting velocity, called the *terminal velocity*. (You will be asked to show this in **Exercise 8**.) However, if it is already known that such a limit exists, it is possible to find the terminal velocity without even solving the differential equation! We can do this by taking the limit as $t \to \infty$ of both sides of the differential equation. At the constant terminal velocity v_T, the left side of the differential equation becomes zero, so that in the case of the linear drag force,

$$0 = -mg - k_1 v_T,$$

$$v_T = -\frac{mg}{k_1}.$$

For the quadratic drag force, we get a terminal velocity of

$$v_T = \sqrt{\frac{mg}{k_2}}.$$

Exercises

5. Use separation of variables to derive **(6)**, and then integrate your solution to get **(7)**.

6. Use separation of variables to derive **(8)** and **(10)**.

7. Integrate **(8)** and **(10)** to get **(9)** and **(11)**.

8. Find the terminal velocity for a falling sphere, that is, evaluate the limit as $t \to \infty$ of $v(t)$, for each of **(6)** and **(10)**.

5.3 Comparison of the Two Models

In a physics class or a calculus class, you probably modeled the vertical motion of a particle by ignoring the drag force. The equation of motion was

$$my''(t) = -mg$$

and the solution was

$$v(t) = -gt + v_0,$$
$$y(t) = -\frac{1}{2}gt^2 + v_0 t + y_0.$$

This model is accurate for objects moving in a vacuum, in which there is no drag force, or sometimes for streamlined objects, such as bullets or rockets, which are designed to have only a small drag force.

Both of our models that include the drag force are refinements of the no-drag model, but they are also more awkward to use. (List some of the extra complications that are involved in using our models.) Our models should be compared to the no-drag model for accuracy. On the other hand, it is meaningless to compare the two drag force models with each other: One is designed to model low Reynolds number flows, and the other is designed to model high Reynolds number flows.

5.4 An Example: Sand Settling in Water

Suppose the Toronto water authority uses Lake Ontario as a water source for the city. After collecting the water, they must allow sand particles to settle. Our problem is to find out how long it will take the sand to settle in holding tanks that are 200 cm deep. The diameter of fine-grained sand is about 5×10^{-3} cm and its density is 2.6 g/cm^3. We must determine if we can use either the linear or the quadratic drag model by checking the size of the Reynolds number. Unfortunately, we must first know the velocity, since $R = \rho v D / \mu$. But we need to know which model to use before we can find the velocity! We will resolve this dilemma by approximating the velocity with the terminal velocity for each model. Then we can check to see if the terminal velocity for either model is consistent with its valid range of Reynolds numbers.

We need to calculate the mass of a sand particle in order to find its terminal velocity. Multiplying the density by the volume, we get

$$m = \left(2.6 \text{ g/cm}^3\right) \times \left(\frac{\pi}{6}D^3\right) = \left(2.6 \text{ g/cm}^3\right) \times \left(\frac{\pi}{6} \times 1.25 \times 10^{-7} \text{ cm}^3\right)$$
$$= 1.70 \times 10^{-7} \text{ g}.$$

For the quadratic drag model, using $C_D \approx 0.5$, the terminal velocity is

$$v_T = \sqrt{\frac{mg}{k_2}} = \sqrt{\frac{mg}{\frac{\pi}{8}\rho D^2 C_D}} = \sqrt{\frac{(1.70 \times 10^{-7})(980)}{\frac{\pi}{8}(1.00)(2.5 \times 10^{-5})(0.5)}}$$
$$= 5.83 \text{ cm/sec}.$$

This gives

$$R = \frac{\rho v_T D}{\mu} = \frac{(1.00 \text{ g/cm}^3)(5.83 \text{ cm/sec})(10^{-2} \text{ cm})}{1.51 \times 10^{-2} \text{ g/cm sec}}$$
$$= 1.93.$$

This is below the appropriate range, $10^3 < R < 10^5$, for the quadratic drag model. Since the velocity of the sand particles will always be below the terminal velocity, we must reject the quadratic drag model for this problem.

Repeating this procedure for the linear drag model, we get

$$v_T = \frac{mg}{k_1} = \frac{mg}{3\pi\mu D} = \frac{(1.70 \times 10^{-7})(980)}{3\pi(1.51 \times 10^{-2})(5.00 \times 10^{-3})}$$
$$= 0.234 \text{ cm/sec},$$
$$R = \frac{\rho v_T D}{\mu} = \frac{(1.00 \text{ g/cm}^3)(0.234 \text{ cm/sec})(10^{-2} \text{ cm})}{1.51 \times 10^{-2} \text{ g/cm sec}}$$
$$= 0.0775.$$

This is well within the range of valid Reynolds numbers, $R < 0.5$, for the linear drag model.

Our problem asks when a particle reaches the bottom of the tank; thus we want to find the value of t when $y = 0$. Substituting $v_0 = 0$ and $y_0 = 200$ into (7) and setting $y = 0$, we get

$$0 = -\frac{mg}{k_1}t - \frac{m^2g}{k_1^2}\left(e^{-\frac{k_1 t}{m}} - 1\right) + 200.$$

Substituting the known quantities for our problem,

$$0 = 0.234t - 5.59 \times 10^{-5}(e^{-4188t} - 1) + 200.$$

This equation cannot be solved explicitly for t, because t appears linearly in the first term and as an exponent in the second. You can approximate the solution—by using Newton's method, or by graphing the function on a graphics calculator or computer and zooming in on the t-intercept. You should get $t = 855$ sec = 14.2 min.

We can approximate the solution by yet another method. Notice that the second term in our equation has a very large negative exponential, which quickly approaches zero. Physically, this means that the particles reach their terminal velocity almost instantaneously. Replacing this term with zero, we get

$$0 = 0.234t + 5.59 \times 10^{-5} + 200 \approx 0.234t + 200,$$

yielding $t = 855$ sec. To check the validity of our approximation, we note that when $t = 855$, the second term is about $10^{-1,555,100}$ —a very small number indeed!

Exercises

9. A room fogger sprays a flea-killing insecticide throughout a room at 20°C with a ceiling 3 meters high. If the droplets range in diameter from 5.0×10^{-4} cm to 2.0×10^{-3} cm, how long should one wait before re-entering the room? Assume the density of the droplets is the same as the density of water. Use the model for low Reynolds number; after solving the problem, verify that the Reynolds number is in the appropriate range.
   Note: We are ignoring two factors which also affect the settling time—the interaction of the droplets, which is significant when the distances between the droplets is small; and the random Brownian motion of the particles, which is significant if the particles are very small ($D < 5 \times 10^{-4}$ cm).

10. Refer to **Exercise 9**. The instructions on a flea bomb state that the room should not be re-entered for at least 2 hours. Predict the size of the droplets.

11. According to an apocryphal tale, Galileo dropped a 100-pound cannonball and a 1-pound ball from the Leaning Tower of Pisa in order to disprove the Aristotelian view that objects fall with velocities proportional to their weights. If the objects hit the ground at the same time, then the Aristotelian school was wrong. Is the no-drag model consistent with Galileo's supposed observation that the balls hit the ground at the same time? Is the drag model? In actuality, Galileo tried a similar experiment in his youth. He dropped a lump of lead and a lump of wood from a high tower. Suppose the tower was 50 m high, the lead was a sphere of diameter 2 cm, the wood was a sphere of diameter 10 cm, and the air temperature was 20°C. The density of lead is 11.3 g/cm³ and the density of wood, say pine, is 0.5 g/cm³. Use the no-drag model to predict the time elapsed before each object hits the ground. Repeat for the drag model.

6. Conclusion

The goal in mathematical modeling is to devise a model that is as simple as possible, yet is accurate enough to make valid predictions of the underlying physical processes. When problems arise that require developing new mathematical models, simple models are usually developed first. The analysis of these simple models can direct the modeler in deciding what refinements must be made to obtain the desired accuracy.

Most modeling involves both theoretical and experimental components. Our study of a sphere moving through a fluid illustrates this phenomenon. We used theoretical analysis—dimensional analysis—in reducing the number of variables. Then we used experimental results to find the relationship between the coefficient of drag and the Reynolds number. This relationship was then incorporated into a theoretical model for particle motion, namely Newton's Second Law. We then used techniques from calculus and differential equations to solve for the velocity and position of the sphere. The final step would be to verify our models from Section 5.2 using experimental measurements of falling spheres.

7. Appendix I: Table of Physical Constants

Densities and viscosities of selected fluids.

Fluid	Density ρ (g/cm^3)	Viscosity μ (g/cm sec)
Dry air at 5°C and 1 atm	1.27×10^{-3}	1.74×10^{-4}
Dry air at 20°C and 1 atm	1.21×10^{-3}	1.81×10^{-4}
Water at 5°C	1.00	1.51×10^{-2}
Water at 20°C	0.998	1.00×10^{-2}
Glycerine	1.26	23.3

8. Appendix II: Dimensional Analysis

Dimensional analysis allows us to determine if the velocity of the sphere, the diameter of the sphere, the density of the fluid, and the viscosity of the fluid affect the drag force only in certain combinations. If some of the variables can be combined, the number of required experiments can be greatly reduced. See Giordano and Weir [1981] or Giordano and Weir [1988] for a more general treatment of this topic.

We must first distinguish between dimensions and units. Dimensions are types of measurements, such as length, mass, velocity, time, and temperature. Length, mass, time, and temperature are fundamental dimensions, while velocity is a derived dimension—it can be expressed as length divided by time. Units are standardized means of labeling dimensions. For example, the dimension length, can be associated with a number of standard units, such as feet, inches, meters, angstrom units, or light years. For a more thorough discussion of dimensions and units, see Strecker [1981].

The objective of dimensional analysis is to combine the variables as products in such a way that the new combined variables are dimensionless. This is possible for any physical law. Physical laws can be written with balanced dimensions; that is, the dimensions of all terms in the equation are the same. It can be shown that any equation with balanced dimensions can be rewritten in terms of dimensionless parameters. Buckingham's Π Theorem tells us how many independent dimensionless parameters we should expect to get. A set of dimensionless parameters is independent if no one of them can be written as a combination of the others. In Buckingham's notation, Π is used for dimensionless parameters.

Buckingham's Π Theorem. *Let n be the total number of dimensional variables in the problem and let r be the maximum number of these variables that will not form a dimensionless product. Then the maximum number of independent dimensionless parameters is $n - r$.*

A weakness of Buckingham's Π Theorem, as written in this form, is that it is difficult to be certain that one has the correct value for r. Usually r equals the number of fundamental dimensions in the problem. This difficulty can be clarified using the concept of rank from linear algebra, but that is beyond the scope of this Module.

Let us see what combinations of our dimensional variables are dimensionless. Let M represent the dimension mass, L represent the dimension length, and T represent the dimension time. Then the dimensions of r, v, D, m, and F_D are M/L^3, L/T, L, M/LT, and ML/T^2, respectively. Our problem has five variables, so $n=5$, and three fundamental dimensions, M, L, and T, so $r=3$. Then Buckingham's Π Theorem states that the dimensional variables can be combined into $n-r = 5 - 3 = 2$ independent, dimensionless parameters; call them Π_1 and Π_2. Any other dimensionless parameters we construct could be expressed in terms of two independent parameters.

We will write our dimensionless parameters as products of powers of the dimensional variables; that is,

$$\Pi_1 = \rho^a v^b D^c \mu^d F_D^e,$$

with a similar equation for Π_2. In order for Π_1 and Π_2 to be independent, we will try to find each one by omitting different factors from the expressions for Π_i. We will omit μ from Π_1 and F_D from Π_2. To obtain Π_1, we will find a, b, c, and e so that

$$\Pi_1 = \rho^a v^b D^c F_D^e$$

is dimensionless. We require

$$\left(\frac{M}{L^3}\right)^a \left(\frac{L}{T}\right)^b L^c \left(\frac{ML}{T^2}\right)^e = M^0 L^0 T^0,$$

which implies that

$$M^{a+e} L^{-3a+b+c+e} T^{-b-2e} = M^0 L^0 T^0.$$

Equating the exponents,

$$a + e = 0$$
$$-3a + b + c + e = 0$$
$$-b - 2e = 0.$$

One solution to this system of 3 linear equations in 4 unknowns is

$$a = -1,\ b = -2,\ c = -2,\ e = 1.$$

This solution yields

$$\Pi_1 = \frac{F_D}{\rho v^2 D^2}.$$

This dimensionless parameter is usually expressed in terms of the cross-sectional area of the sphere, $A = \pi D^2/4$, instead of in terms of D^2. We can replace Π_1 with

$$\Pi_1' = \frac{F_D}{\rho v^2 A}.$$

In **Exercise 14** you will be asked to find Π_2.

Exercises

12. Find another non-zero solution to the system of equations for Π_1.

13. Find an alternative first dimensionless parameter by finding b, c, d, and e such that $\Pi_1 = v^b D^c \mu^d F_D^e$ is dimensionless.

14. Derive $\Pi_2 = \rho v D/\mu$, the Reynolds number, from $\Pi_2 = \rho^a v^b D^c \mu^d$.

9. Sample Exam

1. List three assumptions that are required to use our drag model and give 3 examples of bodies moving through fluids that violate each of the assumptions.

2. The data points in the **Figure 9** show the coefficient of total resistance, C_f, for various values of the Reynolds number, R. The dotted line is a theoretically derived relationship between C_f and R. Find an equation for C_f that approximately satisfies the experimental data.

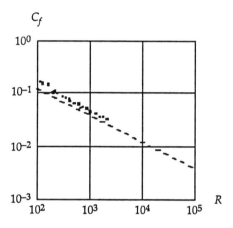

Figure 9. Graph for Problem 2 of the Sample Exam.

3. Suppose you were to construct an experiment to model water flowing past a sphere of diameter 0.1 cm at 100 cm/sec. Your experimental model will use a fluid flowing at 0.2 cm/sec. Would it be more sensible to use glycerine ($\rho = 1.26$ g/cm^3, $\mu = 23.3$ g/cm sec) or water ($\rho = 0.998$ g/cm^3, $\mu = 1.00 \times 10^{-2}$ g/cm sec) in this model? Why? Hint: Recall that $R = \rho v D/\mu$.

4. Use the linear drag force model to find the terminal velocity of a common juniper pollen grain of diameter 3.00×10^{-3} cm, weighing 6.00×10^{-9} g, falling through air at 20°C (for which $\rho = 1.21 \times 10^{-3}$ g/cm^3, $\mu = 1.81 \times 10^{-4}$ g/cm sec). Check to see that the linear drag force model is valid for this problem.

5. A hapless fisherman drops a pearl of diameter 0.5 cm diameter, weighing 0.3 g, into the ocean (for which $\rho = 0.998$ g/cm^3, $\mu = 1.00 \times 10^{-2}$ g/cm sec). How far does it fall in 3 seconds? Use the quadratic drag force model.

10. Solutions to the Exercises

1. $$\frac{(1.26 \text{ g/cm}^3) v (5.00 \text{ cm})}{23.3 \text{ g/cm sec}} = \frac{(0.998 \text{ g/cm}^3)(0.200 \text{ cm/sec})(1.00 \times 10^{-2} \text{ cm})}{1.00 \times 10^{-2} \text{ g/cm sec}}.$$

   Solving for v,
   $$v = \frac{(0.998 \text{ g/cm}^3)(0.200 \text{ cm/sec})(1.00 \times 10^{-2} \text{ cm})}{1.00 \times 10^{-2} \text{ g/cm sec}} \times \frac{23.3 \text{ g/cmsec}}{(1.26 \text{ g/cm}^3)(5.00 \text{ cm})}$$
   $$= 0.738 \text{ cm/sec}.$$

2. The graph for this solution (see **Figure 10**) shows a line close to all of the points except for the first one. This line contains the second point

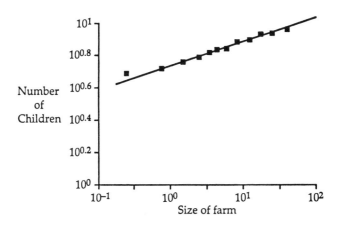

Figure 10. Figure for Solution to Exercise 2.

and the second to the last point. The $\log x$ and $\log y$ coordinates of these points are approximately $(-0.1, 0.7)$ and $(1.4, 0.9)$, respectively. Thus this line has slope $(0.9 - 0.7)/(1.4 - (-0.1)) \approx 0.13$, and its equation is

$$\log y - 0.7 = 0.13(\log x - (-0.1)).$$

Solving for $\log y$, we get

$$\log y = 0.713 + 0.13 \log x.$$

Thus
$$\begin{aligned} y &= 10^{0.713 + 0.13 \log x} \\ &= (10^{0.713})(10^{0.13 \log x}) \\ &= 5.16 \times 10^{(\log x^{0.13})} \\ &= 5.16 x^{0.13}. \end{aligned}$$

21

Using more sophisticated techniques for fitting a line to the data, Styś [1963] obtained $y = 5.51x^{0.14}$.

3. The $\log R$–$\log C_D$ coordinates of the two points in the graph for this solution are approximately $(-1.6, 3)$ and $(0.5, 1)$, so the equation of the line is

$$\log C_D - 1 = \frac{1 - 3}{0.5 - (-1.6)}(\log R - 0.5) = -0.952(\log R - 0.5),$$

$$\log C_D = -0.952(\log R - 0.5) + 1 = -0.952 \log R + 0.476,$$

$$C_D = 10^{-0.952 \log R + 0.476} = 10^{0.476} 10^{-0.952 \log R} = 2.99 \times (10^{\log R})^{-0.952}$$

$$= 2.99 R^{-0.952} = \frac{2.99}{R^{0.952}}.$$

If you picked slightly different coordinates for the two points, your equation for C_D may differ radically from this one. The graph of $24/R$, from Stokes's theory, is shown as a dotted line in **Figure 11**.

4. The horizontal line segment in the graph for this solution is approximately at $C_D = 10^{-0.3} = 0.501$ (see **Figure 12**).

5. Dividing (3) by m, we obtain

$$\frac{dv}{dt} = -g - \frac{k_1}{m}v = w, \quad \text{so that } dw = -\frac{k_1}{m}dv,$$

$$\int \frac{dv}{-g - \frac{k_1}{m}v} = \int dt, \quad -\frac{m}{k_1}\int \frac{dw}{w} = \int dt,$$

$$-\frac{m}{k_1}\ln|w| = t + c, \quad -\frac{m}{k_1}\ln\left|-g - \frac{k_1}{m}v\right| = t + c.$$

Solving for v,

$$\ln\left|-g - \frac{k_1}{m}v\right| = -\frac{k_1}{m}t + c_1,$$

$$\left|-g - \frac{k_1}{m}v\right| = e^{\frac{-k_1}{m}t + c_1} = e^{c_1}e^{-\frac{k_1}{m}t},$$

$$-g - \frac{k_1}{m}v = \pm e^{c_1} e^{\frac{-k_1}{m}t} = c_2 e^{\frac{-k_1}{m}t},$$

$$v = -\frac{mg}{k_1} - \frac{m}{k_1}c_2 e^{\frac{-k_1}{m}t} = -\frac{mg}{k_1} + c_3 e^{\frac{-k_1}{m}t}.$$

Substituting the initial condition that $v = v_0$ when $t = 0$, we get

$$v_0 = -\frac{mg}{k_1} + c_3, \quad c_3 = v_0 + \frac{mg}{k_1}.$$

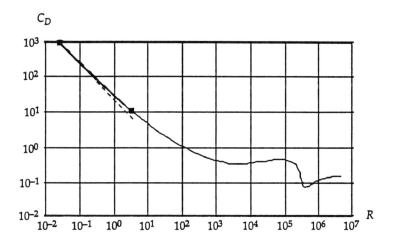

Figure 11. Figure for Solution to Exercise 3.

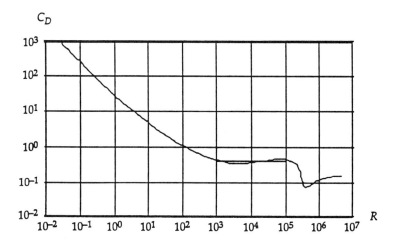

Figure 12. Figure for Solution to Exercise 4.

Substituting this expression for c_3 back into the equation for v, we get

$$v = -\frac{mg}{k_1} + \left(v_0 + \frac{mg}{k_1}\right) e^{\frac{-k_1}{m} t}.$$

Integrating with respect to t, we get

$$y = -\frac{mg}{k_1} t - \frac{m}{k_1}\left(v_0 + \frac{mg}{k_1}\right) e^{\frac{-k_1}{m} t} + c_4.$$

Substituting the initial condition that $y = y_0$ when $t = 0$, we get

$$y = \frac{-m}{k_1}\left(v_0 + \frac{mg}{k_1}\right) + c_4, \qquad c_4 = y_0 + \frac{m}{k_1}\left(v_0 + \frac{mg}{k_1}\right),$$

$$y = \frac{-mg}{k_1} t - \frac{m}{k_1}\left(v_0 + \frac{mg}{k_1}\right) e^{\frac{-k_1}{m} t} + y_0 + \frac{m}{k_1}\left(v_0 + \frac{mg}{k_1}\right)$$

$$= \frac{-mg}{k_1} t - \frac{m}{k_1}\left(v_0 + \frac{mg}{k_1}\right)\left(e^{\frac{-k_1}{m} t} - 1\right) + y_0.$$

6. Applying separation of variables to **(4)**,

$$m\frac{dv}{dt} = -mg - k_2 v^2, \qquad -m\int \frac{dv}{mg + k_2 v^2} = \int dt.$$

This integral is similar to the integral that involves the arctangent. We must get the denominator into the form $1 + u^2$. We can factor mg from the denominator to get

$$\frac{-1}{g}\int \frac{dv}{1 + \frac{k_2}{mg} v^2} = \int dt.$$

We let

$$u = \sqrt{\frac{k_2}{mg}}\, v, \qquad du = \sqrt{\frac{k_2}{mg}}\, dv.$$

Then

$$\frac{-1}{g}\sqrt{\frac{mg}{k_2}}\int \frac{\sqrt{\frac{k_2}{mg}}\, dv}{1 + \frac{k_2}{mg} v^2} = \int dt, \qquad -\sqrt{\frac{m}{k_2 g}}\int \frac{du}{1 + u^2} = \int dt,$$

$$-\sqrt{\frac{m}{k_2 g}} \arctan u = t + c_1, \qquad -\sqrt{\frac{m}{k_2 g}} \arctan\left(\sqrt{\frac{k_2}{mg}}\, v\right) = t + c_1.$$

Solving for v,

$$\arctan\left(\sqrt{\frac{k_2}{mg}}\, v\right) = -\sqrt{\frac{k_2 g}{m}}\, t + c_2, \qquad \left(\sqrt{\frac{k_2}{mg}}\, v\right) = \tan\left(-\sqrt{\frac{k_2 g}{m}}\, t + c_2\right),$$

24

$$v = \sqrt{\frac{mg}{k_2}} \tan\left(-\sqrt{\frac{k_2 g}{m}} t + c_2\right).$$

Substituting the initial condition $v = v_0$ when $t = 0$,

$$v_0 = \sqrt{\frac{mg}{k_2}} \tan c_2, \qquad c_2 = \arctan\left(\sqrt{\frac{k_2}{mg}} v_0\right).$$

Substituting this into the equation for v gives

$$v = \sqrt{\frac{mg}{k_2}} \tan\left(-\sqrt{\frac{k_2 g}{m}} t + \arctan\left(\sqrt{\frac{k_2}{mg}} v_0\right)\right).$$

The solution procedure for **(5)** is analogous to that for **(4)**, except that we must integrate $(1 - u^2)^{-1}$, giving an inverse hyperbolic tangent instead of an inverse tangent.

7. Integrating **(8)** with respect to t to get y,

$$y = \sqrt{\frac{mg}{k_2}} \int \tan\left(-\sqrt{\frac{k_2 g}{m}} t + \arctan\left(\sqrt{\frac{k_2}{mg}} v_0\right)\right) dt.$$

Let

$$w = -\sqrt{\frac{k_2 g}{m}} t + \arctan\left(\sqrt{\frac{k_2}{mg}} v_0\right), \qquad dw = -\sqrt{\frac{k_2 g}{m}} dt.$$

Then

$$\begin{aligned} y &= \frac{-m}{k_2} \int \tan w \, dw = \frac{m}{k_2} \ln|\cos w| + c_3, \\ &= \frac{m}{k_2} \ln\left|\cos\left(-\sqrt{\frac{k_2}{mg}} t + \arctan\left(\sqrt{\frac{k_2 g}{m}} v_0\right)\right)\right| + c_3. \end{aligned}$$

Substituting the initial condition that $y = y_0$ when $t = 0$,

$$y_0 = \frac{m}{k_2} \ln\left|\cos\left(\arctan\left(\sqrt{\frac{k_2}{mg}} v_0\right)\right)\right| + c_3,$$

$$c_3 = y_0 - \frac{m}{k_2} \ln\left|\cos\left(\arctan\left(\sqrt{\frac{k_2}{mg}} v_0\right)\right)\right| = y_0 - \frac{m}{k_2} \ln\left|\frac{1}{\sqrt{1 + \frac{k_2}{mg} v_0^2}}\right|,$$

$$y = \frac{m}{k_2} \ln\left|\cos\left(-\sqrt{\frac{k_2 g}{m}} t + \arctan\left(\sqrt{\frac{k_2}{mg}} v_0\right)\right)\right|$$

$$+ y_0 - \frac{m}{k_2} \ln\left|\cos\left(\arctan\left(\sqrt{\frac{k_2}{mg}} v_0\right)\right)\right|,$$

25

$$= y_0 + \frac{m}{k_2} \ln \left| \frac{\cos\left(-\sqrt{\frac{k_2 g}{m}}\, t + \arctan\left(\sqrt{\frac{k_2}{mg}}\, v_0\right)\right)}{\frac{1}{\sqrt{1+\frac{k_2}{mg} v_0^2}}} \right|$$

$$= y_0 + \frac{m}{k_2} \ln \left| \sqrt{1 + \frac{k_2}{mg} v_0^2}\, \cos\left(-\sqrt{\frac{k_2 g}{m}}\, t + \arctan\left(\sqrt{\frac{k_2}{mg}}\, v_0\right)\right) \right|.$$

The solution procedure for **(11)** is analogous to the above, using the hyperbolic trigonometric identities

$$\int \tanh x\, dx = \ln|\cosh x| + c, \qquad \cosh(\text{arctan} x) = \frac{1}{\sqrt{1-x^2}}.$$

8. For the linear drag force, the terminal velocity is

$$v_T = \lim_{t\to\infty} \left[\frac{-mg}{k_1} + \left(v_0 + \frac{mg}{k_1}\right) e^{\frac{-k_1}{m} t} \right] = \frac{-mg}{k_1}.$$

Eq. **(5)**, for the quadratic drag force, gives a terminal velocity of

$$v_T = \sqrt{\frac{mg}{k_2}}.$$

9. For $D = 5 \times 10^{-4}$,

$$m = \frac{4}{3}\pi (2.5 \times 10^{-4}\text{ cm})^3 (0.988 \text{ g/cm}^3) = 6.53 \times 10^{-11} \text{ g}$$

$$k_1 = 3\pi (1.00 \times 10^{-2} \text{ g/cm sec})(5 \times 10^{-4} \text{ cm}) = 4.71 \times 10^{-5} \text{ g sec},$$

so that

$$v = -1.36 \times 10^{-3} + 1.36 \times 10^{-3} e^{-7.19 \times 10^5 t},$$

$$y = -1.36 \times 10^{-3} t - 1.88 \times 10^{-9} \left(e^{-7.19 \times 10^5 t} - 1\right) + 300.$$

We wish to find t so that $y = 0$, but we cannot solve this equation for t explicitly. We can find an approximate solution by using Newton's method or estimating the t intercept of a computer-generated graph of y. Either method should give $t = 2.21 \times 10^5$ sec = 61.3 hrs.

Alternatively, we can find an approximate value of t by noticing that the negative exponential decays to zero very quickly. Then

$$y = -1.36 \times 10^{-3} t - 1.88 \times 10^{-9} + 300 \approx -1.36 \times 10^{-3} t + 300.$$

The droplet reaches the floor when $-1.36 x 10^{-3} t + 300 = 0$, which implies that $t = 2.21 \times 10^5$ sec = 61.3 hours.

To justify using the linear drag model, we must show that $R < 0.5$.
The velocity is approximately -1.36×10^{-3}, so
$$R = \frac{\rho v D}{\mu} = \frac{(0.998)(1.36 \times 10^{-3})(5 \times 10^{-4})}{1.00 \times 10^{-2}} = 6.78 \times 10^{-5} < 0.5.$$
For $D = 2 \times 10^{-3}$,
$$m = 4.18 \times 10^{-9} \text{ g}, \quad k_1 = 1.88 \times 10^{-4} \text{ g/sec};$$
and
$$v = -0.0217 + 0.0217 e^{-4.50 \times 10^4 t},$$
$$y = -0.0217 t - 4.82 \times 10^{-7} \left(e^{-4.50 \times 10^4 t} - 1 \right) + 300 \approx 0.0217 t + 300,$$
$$t = 1.38 \times 10^4 \text{ sec} = 3.8 \text{ hrs}.$$
The Reynolds number is $R = 4.33 \times 10^{-3} < 0.5$.

If you perform this analysis for spheres with diameters between 5×10^{-4} cm and 2×10^{-2} cm, you will get times between 61 hours and 3.8 hours. Thus, the room will be clear of the pesticide fog in 61 hours.

10. We must show how y depends on D. Since $k_1 = 3\pi\mu D = 9.42 \times 10^{-2} D$ and
$$m = \frac{4}{3}\pi \left(\frac{D}{2}\right)^3 (0.998) = 0.523 D^3,$$
we have
$$y = -5.44 \times 10^3 D^2 t - 5.55 D^2 (v_0 + 5.44 \times 10^3 D^2) \left(e^{-0.180 D^{-2} t} - 1 \right) + 300.$$
Substituting $v_0 = 0$, $t = 7200$ sec, and $y = 0$, we get
$$0 = -3.92 \times 10^7 D^2 - 3.22 \times 10^4 D^4 \left(e^{-1.30 \times 10^3 D^{-2}} - 1 \right) + 300. \quad (12)$$
This equation can be solved approximately by using Newton's method, or by generating a graph of the function on a computer or graphics calculator and estimating the intercept. Alternatively, we can solve it approximately by approximating the negative exponential by zero, getting
$$0 = -3.92 \times 10^7 D^2 + 3.22 \times 10^4 D^4 + 300. \quad (13)$$
Let $x = D^2$. Then (13) becomes a quadratic in x:
$$0 = 3.22 \times 10^4 x^2 - 3.92 \times 10^7 x + 300.$$
Substituting into the quadratic formula, we get
$$x = 7.65 \times 10^{-6} \quad \text{or} \quad x = 1.22 \times 10^3,$$
$$D = 2.77 \times 10^{-3} \quad \text{or} \quad D = 34.9.$$
The second value of D makes the exponent in (12) too large, so that the negative exponential is no longer close to zero (in fact, the Reynolds number is no longer less than 0.5). The first value of D is the only valid solution. Thus, the particles all have diameters greater than or equal to 2.77×10^{-3}.

11. The no-drag model is consistent with Galileo's observation, since the solution of the differential equation is independent of m and D. The drag model predicts that the velocity and distance traveled depend on m and D, so that the balls should have hit the ground at different times.

For the no-drag model, with $v_0 = 0$ and $y_0 = 5,000$, the balls hit the ground when

$$-\frac{1}{2}gt^2 + 5,000 = 0, \quad \text{i.e.,} \quad t = \sqrt{\frac{10,000}{g}} = \sqrt{\frac{10,000}{980}} = 3.19 \text{ sec.}$$

We must now check to see if the linear or quadratic drag force should be used in the drag model. The final velocity using the no-drag model is

$$v = -gt = -(980 \text{ cm/sec}^2)(3.19 \text{ sec}) = -3.13 \times 10^3 \text{ cm/sec.}$$

For the lead ball, this gives a Reynolds number of

$$R = \frac{\rho v D}{\mu} = \frac{(1.21 \times 10^{-3})(3.13 \times 10^3)(2)}{1.81 \times 10^{-4}} = 4.18 \times 10^4.$$

This is well within the range of the quadratic drag force model. Provided that the drag model does not give a radically different descent time from the no-drag model, the quadratic drag model will still have $10^3 < R < 10^5$, and it will thus be self-consistent.

For the quadratic drag model, the balls hit the ground when

$$\frac{-m}{k_2} \ln \left| \cosh \left(-\sqrt{\frac{k_2 g}{m}} t \right) \right| + 5,000 = 0,$$

$$\ln \left| \cosh \left(-\sqrt{\frac{k_2 g}{m}} t \right) \right| = \frac{5,000 k_2}{g},$$

$$\cosh \left(-\sqrt{\frac{k_2 g}{m}} t \right) = e^{5,000 k_2/g}, \quad \cosh \left(\sqrt{\frac{k_2 g}{m}} t \right) = e^{5,000 k_2/g},$$

$$\sqrt{\frac{k_2 g}{m}} t = \operatorname{arccosh} \left(e^{5,000 k_2/g} \right), \quad t = \sqrt{\frac{m}{k_2 g}} \operatorname{arccosh} \left(e^{5,000 k_2/g} \right).$$

For the lead ball,

$$m = \frac{4}{3}\pi(1 \text{ cm})^3(11.3 \text{ g/cm}^3) = 47.3 \text{ g},$$

$$k_2 = \frac{1}{8}\pi \rho D^2 C_D = \frac{1}{8}\pi(1.21 \times 10^{-3} \text{ g/cm}^3)(2 \text{ cm})^2(0.5 \text{ cm}^2/\text{sec}^2)$$
$$= 8.80 \times 10^{-4} \text{ g cm/sec}^2.$$

This gives $t = 3.2$ sec.

For the wooden ball,
$$m = \frac{4}{3}\pi(5 \text{ cm})^3(0.5 \text{ g/cm}^3) = 262 \text{ g},$$
$$k_2 = \frac{1}{8}\pi\rho D^2 C_D = \frac{1}{8}\pi(1.21\times 10^{-3} \text{ g/cm}^3)(10 \text{ cm})^2(0.5 \text{ cm}^2/\text{sec}^2)$$
$$= 2.20\times 10^{-2} \text{ g cm/sec}^2.$$

This gives $t = 3.4$ sec.

12.
$$a + d = 0$$
$$-3a + b + c + d = 0$$
$$-b - 2d = 0.$$

If we set $d = 2$, we get $a = -2$ from the first equation and $b = -4$ from the third equation. Substituting into the second equation gives $c = -4$. This gives the dimensionless parameter, $\rho^{-2}v^{-4}D^{-4}F_D^2$, which is the square of our original Π_1. Other choices for d would give Π_1 to other powers.

13. The units for this new Π_1 are
$$\left(\frac{L}{T}\right)^b L^c \left(\frac{M}{LT}\right)^d \left(\frac{ML}{T^2}\right)^e = M^{d+e}L^{b+c-d+e}T^{-b-d-2e}.$$

Setting this equal to $M^0L^0T^0$ and equating exponents gives
$$d + e = 0$$
$$b + c - d + e = 0$$
$$-b - d - 2e = 0.$$

We have three equations in four unknowns, so we may choose the value of one unknown arbitrarily. Let $e = 1$; we then solve for the other three unknowns. The first equation gives $d = -1$, substituting into the third equation gives $b = -1$, and substituting into the second equation gives $c = -1$. Then
$$\Pi_1 = v^b D^c \mu^d F_D^e = v^{-1}D^{-1}\mu^{-1}F_D^1 = \frac{F_D}{\mu v D}.$$

A different choice for e would have given this quantity raised to a power.

14. The units for Π_2 are
$$\left(\frac{M}{L^3}\right)^a \left(\frac{L}{T}\right)^b L^c \left(\frac{M}{LT}\right)^d = M^{a+d}L^{-3a+b+c-d}T^{-b-d}.$$

Setting this equal to $M^0L^0T^0$ and equating exponents gives
$$a + d = 0$$
$$-3a + b + c - d = 0$$
$$-b - d = 0.$$

As in **Exercise 13**, we have three equations in four unknowns, so we may choose the value of one unknown arbitrarily. Letting $d = 1$, we get $a = -1$ from the first equation, $b = -1$ from the third, and $c = -1$. This gives

$$\Pi_2 = \rho^a v^b D^c \mu^d = \rho^{-1} v^{-1} D^{-1} \mu^1 = \frac{\mu}{\rho v D}.$$

This is the reciprocal of what we wanted, so we should have chosen d to be -1, which would give $a = b = c = 1$.

11. Answers to the Sample Exam

1. See the text of this Module, pp 1–2.
2. $C_f = 0.158/\sqrt{R}$.
3. Glycerine, because the sphere would have diameter 9.25 cm. Using water would require a very small sphere, with diameter 0.005 cm.
4. $v_T = 1.15$ cm/sec, $R = 0.0230 < 0.5$.
5. 228 cm.

References

Diamond, Jared. 1989. How cats survive falls from New York skyscrapers. *Natural History* (August 1989): 20–26.

Giordano, Frank, and Maurice Weir. 1981. Dimensional analysis. UMAP Modules in Undergraduate Mathematics and Its Applications: Module 526. *The UMAP Journal* 2(3)(1981): 97–123.

_____. 1988. *A First Course in Mathematical Modeling.* Pacific Grove, CA: Brooks-Cole.

Millikan, R.A. 1963. *The Electron.* Phoenix Science Series, PSS523. Chicago, IL: University of Chicago Press.

Purcell, E.M. 1977. Life at low Reynolds numbers. *American Journal of Physics* 45(1): 3–11.

Strecker, George E. 1981. Keeping dimensions straight. UMAP Modules in Undergraduate Mathematics and Its Applications: Module 564. Newton, MA: Educational Development Center, Inc. (available from COMAP, Inc.)

Styś, W. 1963. Zludzenia statystyczne wywolane przez wplyw czasu w badaniach zjawisk demograficznych w ich ruchu i rozwoju. Prezeglad Antropologiczny, v. 23. Cited in *Linear Regression and Its Applications to Economics*, by Zdzislaw Hellwig, translated from the Polish by J. Stadler (Oxford, England: Pergamon).

About the Author

H. Edward Donley received his B.A. in mathematics from Grove City College and his M.S. and Ph.D. degrees in mathematics from Carnegie Mellon University. His major mathematical interests are numerical analysis, fluid mechanics, and parallel algorithms. His non-mathematical interests include canoeing, hiking, woodworking, guitar, and banjo.

UMAP

Modules in Undergraduate Mathematics and its Applications

Published in cooperation with the Society for Industrial and Applied Mathematics, the Mathematical Association of America, the National Council of Teachers of Mathematics, the American Mathematical Association of Two-Year Colleges, The Institute of Management Sciences, and the American Statistical Association.

Module 714

Heat Therapy for Tumors

Leah Edelstein–Keshet

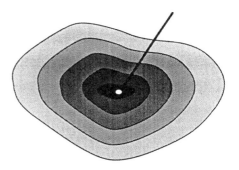

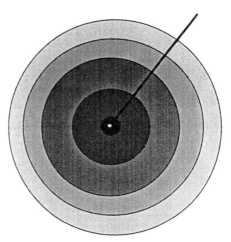

Applications of Calculus to Medicine

72 Tools for Teaching 1991

INTERMODULAR DESCRIPTION SHEET:	UMAP Unit 714
TITLE:	Heat Therapy for Tumors
AUTHOR:	Leah Edelstein-Keshet Mathematics Dept. University of British Columbia Vancouver, B.C. Canada V6T 1Z2
MATHEMATICAL FIELD:	Calculus
APPLICATION FIELD:	Medicine
TARGET AUDIENCE:	Third-semester calculus students.
ABSTRACT:	Calculus techniques are used to gain insight into a problem of clinical and medical significance. The methods used include triple (volume) integration in spherical coordinates and elementary Taylor series. They are applied to evaluating *hyperthermia*, a form of cancer therapy that has gained recognition and importance in the last decade.
PREREQUISITES:	Understanding of equithermal surfaces and of spherical and cylindrical coordinates; and exposure to triple integration in spherical coordinates and to Taylor series.

©Copyright 1991, 1992 by COMAP, Inc. All rights reserved.

COMAP, Inc., Suite 210, 57 Bedford Street, Lexington, MA 02173
(800) 77–COMAP = (800) 772–6627,   (617) 862–7878

Heat Therapy for Tumors

Leah Edelstein-Keshet
Mathematics Dept.
University of British Columbia
Vancouver, B.C.
Canada V6T 1Z2

Table of Contents

1. INTRODUCTION 1
2. HYPERTHERMIA 1
3. MONITORING THE COURSE OF A TREATMENT 2
4. CHARACTERIZING TUMOR TEMPERATURES 4
5. MODELING THE PROCESS AND MAKING SIMPLIFYING ASSUMPTIONS . . . 4
6. TUMOR GEOMETRY AS THE DOMINANT EFFECT 8
   6.1 Spherical Tumor, Linear Temperature Profile 9
   6.2 Nonspherical Tumor, Monotonic Temperature Profile 11
7. HEATING PATTERN AS THE DOMINANT EFFECT 14
   7.1 Some Equitherms Are (Almost) Ellipsoidal 17
   7.2 Some Equitherms Are (Almost) Parabolic 18
   7.3 Estimating Heated Volumes 19
8. EPILOGUE 21
9. SOLUTIONS TO THE EXERCISES 23
   REFERENCES 27
   ACKNOWLEDGMENTS 27
   ABOUT THE AUTHOR 28

MODULES AND MONOGRAPHS IN UNDERGRADUATE
MATHEMATICS AND ITS APPLICATIONS (UMAP) PROJECT

The goal of UMAP is to develop, through a community of users and developers, a system of instructional modules in undergraduate mathematics and its applications, to supplement existing courses and from which complete courses may be built.

The Project was initially funded by a grant from the National Science Foundation and has been guided by a National Advisory Board of mathematicians, scientists, and educators. UMAP is now supported by the Consortium for Mathematics and Its Applications (COMAP), Inc., a non-profit corporation engaged in research and development in mathematics education.

COMAP Staff

Paul J. Campbell	Editor
Solomon Garfunkel	Executive Director, COMAP
Laurie W. Aragón	Development Director
Philip A. McGaw	Production Manager
Roland Cheyney	Project Manager
Laurie M. Holbrook	Copy Editor
Dale Horn	Design Assistant
Rob Altomonte	Distribution Coordinator
Sharon McNulty	Executive Assistant

1. Introduction

Tumors that have not responded to conventional surgery and chemotherapy can sometimes be affected by the relatively new treatment *hyperthermia*, which may provide remission or partial arrest of tumor growth.

In hyperthermia treatment, a tumor is heated by microwaves applied from outside the body. The internal temperature of the tumor must be monitored using probes in catheters embedded in the tissue. How can a small number of measurement samples be used to assess and evaluate the course of therapy? We use geometry and simple concepts from calculus to estimate the fraction of a tumor that has achieved temperatures in a therapeutic range, based on temperature measurements along one or two catheters.

2. Hyperthermia

The premise underlying hyperthermia is that living cells, whether normal or malignant, are adversely affected by temperatures beyond some normal tolerance level. Normal tissues are protected from heat damage by an elaborate physiological response in which blood vessels dilate locally, allowing blood flow to flush heat away. This mechanism operates efficiently (within limits) in any tissue with a normal density of blood vessels. Tumors, however, while surrounded by a thick density of vessels on their periphery, are notoriously poor in blood vessels in their interior. This fact can be exploited by heating the tumor and some normal surrounding tissue to temperatures above a therapeutic level at which tumor cells start to die (e.g., 43–45°C). Generally, the interior core of the tumor is more easily brought to the desired temperatures than its outer regions, which may remain at relatively normal body temperatures or at insignificantly elevated ones.

For a tumor close to the surface of the body, a method of heating popular in clinical applications uses microwave applicators outside the body (see **Figure 1** for a typical set-up). The heat is applied for a period of about 30 minutes, repeated several times weekly.

However, one drawback and nuisance factor persists in this apparently simple idea: It is currently impossible, using external means alone, to monitor the "dose" of heat applied, or the actual temperatures achieved within the tumor. Moreover, individual properties of tumors—their geometry, the normal tissue environment, the arrangement of blood vessels, and the way these change throughout the treatment—differ from one case to another, making it a challenge to extrapolate from one individual to the next.

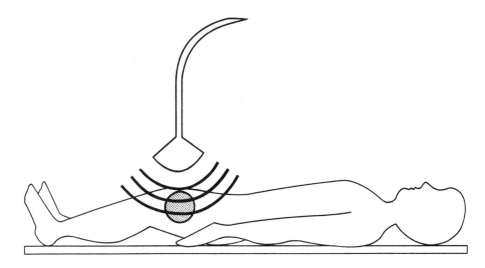

Figure 1. A patient with a tumor in the thigh being treated by hyperthermia.

3. Monitoring the Course of a Treatment

For the reasons outlined above, hyperthermia treatment must be monitored invasively, using probes implanted into the tumor to record internal tumor temperatures (see **Figure 2a**). These probes may cause discomfort and anxiety in patients, but this may be a small price to pay for a possibly lifesaving remedy.

Current technology permits multiple temperature measurements along a single line segment, the length of a catheter embedded into the tumor. **Figure 2a** illustrates a typical placement of such a thermometry catheter. A temperature-sensitive probe is translated along the catheter in precise computer-controlled steps.

The temperature distribution measured along such a probe is shown in **Figure 2b**. We observe that the temperature is highest in the core of the tumor and lowest at the skin surface. **Figure 3a** gives a schematic diagram of another tumor-catheter configuration, in which the catheter spans the radius of the tumor, rather than its diameter. (The hypothetical temperature distribution in **Figure 3b** is discussed in **Exercise 1**).

Because of the undesirable physical and psychological effects of inserting many probes into the tumor region, the goal of this unit (and the medical paper [Edelstein-Keshet et al. 1989] upon which it is based) is to extract maximal information about tumor temperatures from a minimum number of measurements. A subsidiary goal is to help guide a clinician about where catheters should optimally be placed and how the data so gathered might be used.

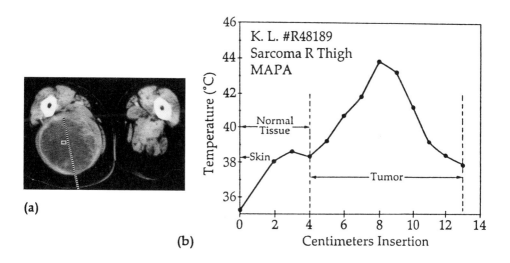

Figure 2a. Tomography scan of a cross-section of a patient's thighs reveals a large spherical tumor called a *sarcoma* (left image). A stippled line represents the location of a temperature probe. **Figure 2b.** Temperatures measured along this probe during a hyperthermia treatment. The tumor attains temperatures higher than those of surrounding tissue. Since the temperature distribution is strongly influenced in this case by the geometry of the tumor, the approximations of Section 6 apply to this example.

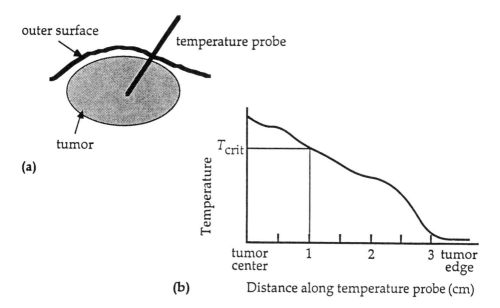

Figure 3a. The placement of a temperature probe. **Figure 3b.** The temperature profile along the probe. One-third of the probe registers temperatures above the critical (therapeutic) value.

4. Characterizing Tumor Temperatures

Trying to ascertain the exact internal temperature distribution of a tumor is both hopeless and unnecessary. Instead, some simpler function of the temperature may be of much greater clinical value. Such quantities have been called *descriptors*.

What constitutes an adequate descriptor continues to be a controversy in medical circles. Here we shall focus on a simple example, the relative volume of the tumor heated above some critical temperature T. We denote this quantity by V_T/V, where V is tumor volume and V_T is the volume heated above temperature T.

We will make some simplifying assumptions and treat two extreme cases, each with a similar underlying question, paraphrased as follows: Given temperature measurements along one (or possibly two) lines through the tumor, what can we conclude about the fraction of volume that achieves therapeutic temperatures?

Exercise

1. **Figure 3b** gives a temperature distribution obtained during a typical hyperthermia session. A clinical technician observes that approximately one-third of the probe has attained temperatures above T_{crit}, where true therapeutic effects are to be expected. The technician concludes that one-third of the tumor has been heated successfully. Comment on this conclusion.

5. Modeling the Process and Making Simplifying Assumptions

One approach to the tumor temperature problem is to solve a partial differential equation governing heat distribution, often called the *bioheat equation*:

$$\frac{\partial T}{\partial t}(\vec{x},t) = k_t \nabla^2 T(\vec{x},t) + \dot{Q}(\vec{x},t) - w_b c_b T(\vec{x},t), \tag{1}$$

where

$T(\vec{x},t)$ is the temperature elevation above the arterial baseline temperature,

k_t is the thermal conductivity,

w_b is a blood-flow heat-transfer parameter,

c_b is the heat capacity of blood, and

$\dot{Q}(\vec{x},t)$ is the absorbed-power density distribution resulting from the heat applicator.

The equation describes how the temperature changes in a three-dimensional region when external heating and internal cooling takes place. It is based on the physical principle that heat is conserved, so that sources and dissipation of heat must balance out.

However, we shall not use this approach, because in most clinical settings it proves to be too complex. The difficulties stem from

- the unknown spatial variation in k_t due to different types and properties of tissues surrounding the tumor;

- the unknown spatial-temporal variation in the blood-flow distribution pattern; and

- the geometric complexity of boundaries, surfaces, and heating patterns.

Such full thermal models are nonetheless in use (see the References). They rely extensively on numerical and computational techniques beyond the scope of this article. Rather than dwell on these advanced techniques, we invoke simple mathematical and geometric arguments to gain some understanding of the conversion of one-dimensional data (temperatures along line segments) to statements about volumes.

To do so, first recall that, because of the pattern of blood flow around a tumor, tumor temperature patterns are generally hottest *inside* and cooler toward the periphery. Suppose we consider all points with the same temperature T_0, i.e., points (x, y, z) such that

$$T(x, y, z) = T_0.$$

Such a set of points is commonly called an *equitherm*, an *isotherm*, or an *equitemperature contour*. In three dimensions, such a set is a surface.

We shall generally assume that there is a single point (the "center" of the tumor) at which some maximal temperature is attained. This means that the equitherms will be closed surfaces nested one inside the next, growing progressively cooler toward the outside.

In the following sections we will find it useful to express equithermal suraces in spherical or cylindrical coordinates. Recall that the equation of a sphere, and thus the equation describing the periphery of a spherical tumor, is simplest in spherical coordinates:

$$\rho = R = \text{constant},$$

where R is the radius of the sphere and ρ is the distance of any point in space from the origin (at the center of the sphere). See **Figure 4** for descriptions of the spherical and cylindrical coordinate systems.

In order to make some progress on our problem, we must make certain simplifying assumptions. Below we consider two distinct cases, each one an extreme possibility.

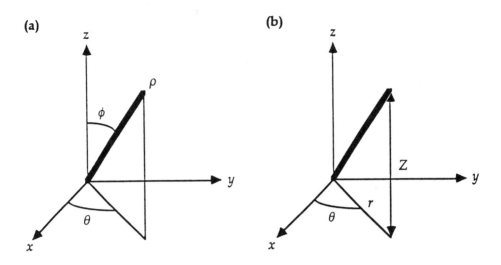

Figure 4a. Spherical coordinates. **Figure 4b.** Cylindrical coordinates.

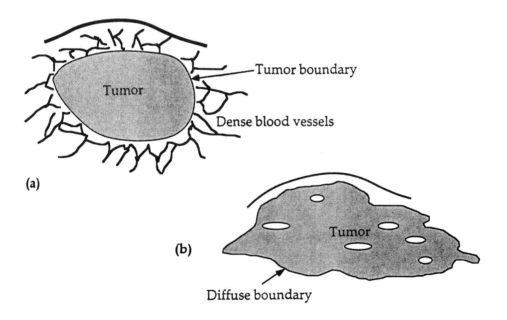

Figure 5. Two extreme possibilities. **a.** The tumor boundary is well-defined, and the geometry of the tumor is the chief factor governing the temperature distribution. **b.** The tumor is diffuse and nearly uniform, and the properties of the heat applicator are assumed to predominate in determining the temperature distribution.

In the first case (Section 6), we shall assume that the shapes of the equitherms are most keenly affected by the geometry of the tumor. That is to say, the boundary of the tumor has such distinct characteristics that it dominates in shaping the temperature profiles. This would be true if, for example, the tumor periphery had numerous blood vessels that efficiently maintain a fixed temperature all round the periphery (see **Figure 5a**). This is also evidently the case in the patient whose tumor is shown in **Figure 2**, since the temperature distribution of **Figure 2a** clearly changes abruptly at the tumor edge.

In the second case (Section 7), we consider a diffuse malignant area lacking a well-defined border, which has temperature distributions "borrowed" from properties of the heating deposition pattern of the applicator (see **Figure 5b**).

It is important to keep in mind that these extremes are simplified special cases, neither of which exactly addresses the full complexity of a realistic case. Such shortcuts and simplifications are nevertheless a key part of any initial modeling process in which some intuition is gained. The fact that simple calculus suffices in these calculations make these two case studies excellent examples of undergraduate mathematics that has realistic applications.

Exercise

2. Draw equithermal surfaces for each of the following temperature distributions.

   a) $T(x, y, z) = T_{max} - \left(\dfrac{x^2}{a^2} + \dfrac{y^2}{a^2} + \dfrac{z^2}{b^2}\right),$

   b) $T(x, y, z) = T_{max} - (x^2 + y^2 + z^2),$

   c) $T(\rho, \theta, \phi) = T_{max} - \rho,$

   d) $T(\rho, \theta, \phi) = T_{max} - \dfrac{\rho}{1 + 0.2 \sin 8\theta \sin \phi},$

   e) $T(\rho, \theta, \phi) = T_{max} - \dfrac{\rho}{1 + 0.2 \sin 8\theta \sin 4\phi}.$

6. Tumor Geometry as the Dominant Effect

In our first limiting case, we assume that the temperature of the tumor periphery is always maintained at some fixed baseline temperature, T_{edge}, by the rapid cooling effect of the surrounding blood flow. Note that this will constrain the shape of the equitherm $T(x, y, z) = T_{\text{edge}}$ to be identical with the shape of the tumor.

A second assumption is that there is a single temperature maximum somewhere inside the tumor. We refer to this site as the "center" of the tumor, although in principle it need not coincide with the center of mass. We consider the situation in which temperature measurements are known to any degree of accuracy along one radius connecting the center and the periphery, i.e., a single catheter with a temperature probe spans the tumor radius (see **Figures 2, 3, and 6**).

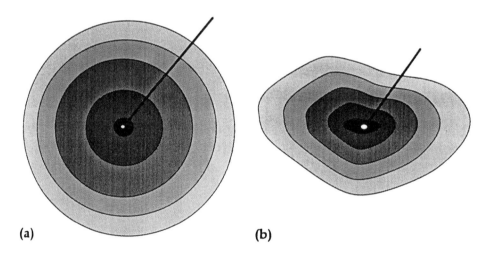

Figure 6. Tumors for which the boundary is the most important determinant of the shape of the equitemperature contours. **a.** A spherical tumor. **b.** A nonspherical tumor.

The goal of the section is to estimate the ratio of V_T/V from these data. We proceed from the simplest case, a spherical tumor with a linear temperature gradient, to a slightly more elaborate one, in which the tumor geometry is more complicated and the temperature nonlinear.

6.1 Spherical Tumor, Linear Temperature Profile

We will take the center of the tumor as the origin of our spherical coordinate system. Suppose that the maximal temperature (at the center of the tumor) is $T_{\max}$ and the minimum (along the periphery) is T_{edge}. We shall use α to denote a relative distance along the tumor radius, so that

$$\alpha = \frac{\rho}{R}. \tag{2}$$

Observe that for points on the inside of the sphere, we have $0 < \alpha < 1$, and that (2) implies that $\rho = \alpha R$. Thus, distances can be measured in fractional units of the tumor radius.

Suppose we observe that the temperature is linear along a temperature probe. Then, in terms of α, we have

$$T(\alpha) = T_{\max} - \alpha(T_{\max} - T_{\text{edge}}), \quad 0 < \alpha < 1. \tag{3}$$

Note that at $\alpha = 0$, $T = T_{\max}$, and at $\alpha = 1$, $T = T_{\text{edge}}$. But these data are incomplete. We have no information about temperatures at other sites. Is there a simple, logical way of extending this one-dimensional data set to fill a three-dimensional region? One way would be to assume that all directions are equivalent. In that case, the value of temperature at any point (in spherical coordinates) (ρ, θ, ϕ) can be estimated.

Our goal is to determine V_T/V, the fraction of volume above some critical temperature T_{crit}. The problem is to calculate the volume V_T, since we know that the tumor volume is $V = \frac{4}{3}\pi R^3$. Observe that V_T is simply the volume enclosed by the equithermal surface $T = T_{\text{crit}}$. Thus, the problem reduces to finding the size of a region bounded by an equithermal surface. **Exercise 4** shows that we can compute the ratio V_T/V from just T_{edge}, $T_{\max}$, and the desired critical value of T.

Exercises

3. Argue that the above assumption of spherical symmetry implies that the temperature distribution in spherical coordinates is

$$T(\rho) = T_{\max} - m\rho,$$

where

$$m = \frac{T_{\max} - T_{\text{edge}}}{R}$$

is the slope of the temperature gradient.

4. Show that (3) implies that the equithermal surfaces satisfy an equation of the form

$$\rho = \alpha_T R, \tag{4}$$

where

$$\alpha_T = \frac{T_{max} - T(\alpha)}{T_{max} - T_{edge}}. \tag{5}$$

Now find that the ratio V_T/V is thus simply the ratio of $\frac{4}{3}\pi\alpha_T^3 R^3$ to $\frac{4}{3}\pi R^3$, namely,

$$\frac{V_T}{V} = \alpha_T^3.$$

5. Consider the linear temperature profile shown in **Figure 7** for a hypothetical tumor of radius 2.0 cm. Suppose the critical temperature for therapeutic effect is 45°C. Estimate the fraction of the tumor that has received a therapeutic heat dose. (Hint: First find α_T, the fraction of the temperature probe which registers temperatures in excess of T_{crit}.)

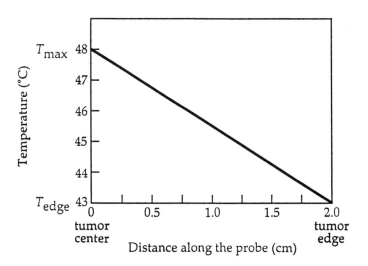

Figure 7. A hypothetical linear temperature profile, as measured along the temperature probe shown in **Figure 6a**.

6.2 Nonspherical Tumor, Monotonic Temperature Profile

While the above simple results may appear to depend intimately on the properties of a sphere, we can generalize them considerably to include special classes of nonspherical shapes and nonlinear temperature profiles.

We shall assume, as before, that there is a unique point inside the tumor at which a temperature maximum T_{max} is attained, and that the peripheral temperature T_{edge} is fixed.

Now, however, we allow the shape of the tumor to be more general. Indeed, we consider any shape whose edge is a surface that satisfies an equation of the form

$$\rho = R(\theta, \phi), \qquad (6)$$

where θ and ϕ are the angles in spherical coordinates, as before.

What kind of shapes could fit this description? Spheres, bumpy and distorted spheres, and "star-shaped" surfaces are all members of this class (see **Figure 6b** and **Exercise 2d** and **2e** for examples). Eq. (6) says that a ray emanating from the origin along direction θ, ϕ intersects the surface at exactly one location, whose distance from the origin is R. This description precludes surfaces that bend back on themselves or have spiralling appendages but leaves scope for a considerable variety of realistic tumor shapes.

Suppose a temperature probe spans the radius of the tumor along a direction specified by the angles θ, ϕ. (We assume that one endpoint of the probe is at the tumor "center," which corresponds to the origin of the coordinate system.) A physical distance ρ along this catheter would correspond to some fraction α of the radius of the tumor in that particular direction:

$$\alpha(\rho, \theta, \phi) = \frac{\rho}{R(\theta, \phi)}.$$

The quantity α will again be used to compare temperature behavior along different radii of the tumor, whose true lengths need no longer be the same.

This time we treat the case in which the observed temperature data are not assumed to be linear, i.e., the temperature T at a fraction of the distance α along the probe satisfies

$$T(\alpha) = f(\alpha), \qquad (7)$$

where f is any strictly decreasing continuous function of α with $f(1) = T_{edge}$ and $f(0) = T_{max}$ (the maximal temperature, which occurs at the "center"). Note that this equation encompasses many possible functional forms, provided there are no local maxima or minima in the interval $(0,1)$. We require f to be strictly decreasing, since we are then assured that it has an inverse,

$$\alpha_T = f^{-1}(T). \qquad (8)$$

Further, the fact that f is continuous allows us to conclude that every temperature value in the range $T_{\text{edge}} < T < T_{\max}$ is attained at some unique fractional distance α along the temperature probe.

The one-dimensional measurement must again be extended to a full three-dimensional temperature distribution. We do this by making an assumption similar to the one made in the previous spherically symmetric case, namely, that:

Every direction in the tumor has an equivalent temperature profile, provided the radius is scaled so that the center occurs at $\alpha = 0$ and the tumor edge at $\alpha = 1$.

This assumption means that the temperature at any point (ρ, θ, ϕ) depends only on the distance of the point from the origin, ρ, scaled by the radius $R(\theta, \varphi)$, i.e.,

$$T(\rho, \theta, \phi) = f\left(\frac{\rho}{R(\theta, \phi)}\right) = f(\alpha). \tag{9}$$

Is such an assumption physically plausible? For tumors close to spherical, with a single temperature maximum and constant edge temperature, the approximation *is* an accurate one. For shapes deviating significantly from a sphere, equitherms close to the center would be smoother than a reduced copy of the tumor boundary, because heat conduction smooths gradients fastest at small distances. Nevertheless, this assumption is a simple natural extension of the spherically symmetric case, which serves well as an initial approximation.

As a final step, in **Exercise 7** we compute the volume of the equitherm $T(\rho, \theta, \varphi) = T_{\text{crit}}$. This can be done by a straightforward triple integration, which is best done in spherical coordinates.

Exercises

6. Show that the assumption of equivalent temperature profiles in every direction implies that equithermal surfaces are described by equations of the form

   $$\rho = \alpha_T R(\theta, \varphi),$$

   where

   $$\alpha_T = f^{-1}(T), \quad 0 \le \alpha_T \le 1; \tag{10}$$

   i.e., these surfaces are copies of the tumor boundary contracted about its center (see **Figure 6b**), as though similarly shaped balloons were placed one inside the other.

7. Find the volume inside an equitherm and show that

$$V_T/V = \alpha_T^3.$$

Note that the integration can be done, even though the function f in **(7–9)** is not specified. Observe that this result agrees with previous findings, and conclude that our estimate for the fraction of the region heated above a given temperature is

$$V_T/V = [f^{-1}(T)]^3.$$

8. Consider the nonlinear temperature profile shown in **Figure 8** and obtained as in **Exercise 5**. Give a numerical estimate for the ratio V_T/V.

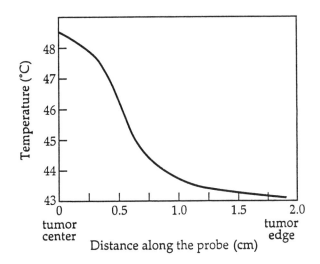

Figure 8. A temperature profile, measured across the length of a probe, inside a tumor with the geometry shown in **Figure 5b**. In this hypothetical example, $T_{\max} = 48.5°C$ and $T_{\text{edge}} = 43°C$.

7. Heating Pattern as the Dominant Effect

In a second extreme possibility, we assume that the malignant region is fairly uniform, and the temperature distribution is dominated by the heat deposition pattern rather than by the tumor geometry. We focus on a standard heating pattern and attempt to estimate the volume of tissue heated beyond some critical temperature. As we soon find out, an exact functional description of the temperature at all points of the region is not always easily converted to a statement about volumes inside equitherms. However, some judicious approximations will lead us to quick ways of overcoming the problem without the necessity of performing nasty volume integration.

Figure 9 demonstrates a typical configuration of microwave heat applicator, skin surface, bag containing coolant applied to the skin (to prevent surface burns), and region to be treated. Surface cooling can maintain the skin at a fixed baseline. Moreover, little heat energy is deposited deep into the tissue, so normal body temperature persists there.

One type of microwave applicator, with a round disk-like shape, is known to produce an absorbed energy density distribution that falls off exponentially with depth into the tissue. Since this applicator is cylindrically symmetric, it is convenient to use cylindrical coordinates (z, r, θ) (see **Figure 6b**), and we shall take the convention that z is depth into the tissue ($z = 0$ at the skin surface) and r is the distance from the vertical axis of the applicator (see **Figure 9**). The absorbed energy density distribution of this applicator

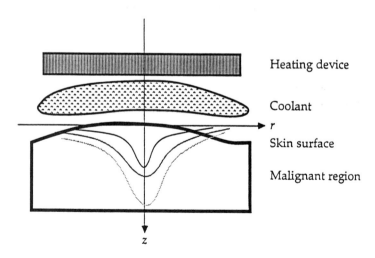

Figure 9. A diffuse malignant region, with a temperature distribution influenced predominantly by the position and the properties of the heating device. The superimposed curves represent the intensity of heating (see **(11)**. Note that the positive z-axis represents depth into the region.

(according to design specifications) is usually of the form

$$Q(r,z) = Q_0 \exp(-\mu z) \exp\left(-(\sigma r)^2\right), \tag{11}$$

where

μ is the linear attenuation coefficient in tissue,

σ defines the width of the distribution in the radial direction, and

Q_0 is the power density at $(0,0,0)$.

The equation simply means that the heat energy delivered by the applicator falls off exponentially with distance into the tissue, and it also decays radially away from the applicator along the skin surface in a cylindrically symmetric "Gaussian" shape. The most heat is thus delivered along the z-axis, close to the skin.

To obtain a mathematical description of the steady-state tissue temperature distribution that results from this heating pattern, one needs to solve a steady-state version of (1) (taking $\partial T/\partial t = 0$) that also takes into account the cylindrical symmetry. Solutions of heat equations often involve the technique of *separation of variables*, taught in introductory courses on differential equations. In cylindrically symmetric cases, the ordinary differential equations that result from this procedure have special types of solutions, such as Bessel functions; but the "routine" applied mathematics task involved is well beyond our scope. Anyway, the rather unpleasant solutions reduce to a conveniently simple form, provided one assumes that blood flow rapidly dissipates heat. It can then be shown that the temperature at any point (r, z, θ) can be approximated by

$$T(r,z) = A \exp\left(-(\sigma r)^2\right) \left[\exp(-\mu z) - \exp\left(-\omega^{1/2} z\right)\right], \tag{12}$$

where

$$A = \frac{Q_0}{k_t(\omega - \mu^2)},$$

and

$$\omega = \frac{\omega_b c_b}{k_t},$$

with ω the ratio of blood cooling properties to thermal conductivity, σ and μ properties of the heating applicator described in (11), and the other parameters are as defined in (1). The fact that this solution is a good approximation for $\omega^{1/2} > \mu$ is discussed in an advanced paper by Samulski et al. [1988].

As before, our goal is to estimate the volume V_T of the region heated to above some critical temperature T_{crit}. Since an estimate of the temperature distribution is explicitly available in (12), we hope that the task should be a relatively easy one. A first step is to find the equations describing equithermal surfaces $T(r, z, \theta) = T_{\text{crit}}$, then to determine their volume by some method.

Exercise 9b demonstrates that an explicit formula for the temperature distribution is not always readily convertible to an estimate for the quantity V_T, especially when complicated functions are to be integrated. We turn instead to some revealing "rough estimates." The cross-sections of the equitherms given by **(13)** and shown in **Figure 10** have the following property:

Close to the maximum, the cross-sections of the equitherms resemble ellipses, whereas at great depths they take a more parabolic appearance.

We can use an elementary "trick" involving a Taylor-series computation to verify this hunch! Further, we will exploit this fact in designing a simple way to estimate V_T.

Exercise

9. a) Show that equitherms to **(12)** are given by

$$r^2(z) = \frac{1}{\sigma^2}\left(-\mu z + \ln\left[1 - \exp\left(\mu - \omega^{1/2}\right)z\right] + C\right), \tag{13}$$

where $C = \ln(A/T_{\text{crit}})$. Reason that these equitherms are *surfaces of revolution*, i.e., have cylindrical symmetry. A family of such surfaces is shown (in cross-section) in **Figure 10**.

b) Set up an integral for the volume inside the above surfaces of revolution. (Hint: Use either the disk method or the shell method of integration.) Can this integral be easily evaluated?

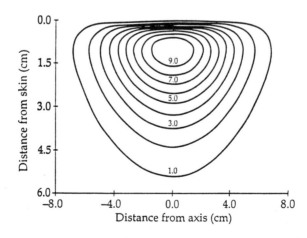

Figure 10. Cross-sections of the equitherms given by **(13)**, shown in the zr-plane. These equitherms are surfaces of revolution, with the z-axis as the axis of symmetry. The skin surface is at the top, and the positive z-axis points down, as in **Figure 9**. The numbers on the equitherms represent the temperatures above baseline, in degrees Celsius. [Adapted from Edelstein-Keshet et al. [1989], Figure 3.]

7.1 Some Equitherms Are (Almost) Ellipsoidal

We observe that the equations of equitherms are of the form

$$\frac{T_{\text{crit}}}{A} R(r) = F(z), \qquad (14)$$

where

$$F(z) = \exp(-\mu z) - \exp(-\omega^{1/2} z),$$
$$R(r) = \exp\left(-(\sigma r)^2\right).$$

Exercises

10. a) Show that the maximum temperature occurs at a depth

$$z^* = \frac{\ln \gamma}{\omega^{1/2} - \mu} = \frac{\ln \gamma}{\mu(\gamma - 1)}, \qquad (15)$$

where $\gamma = \omega^{1/2}/\mu$. (Observe that $\gamma > 1$, since $\omega^{1/2} > \mu$ by assumption.)

b) Show that the maximum temperature, attained at z^*, is

$$T_{\max} = T(z^*) = A e^{-\mu z^*} (1 - 1/\gamma).$$

11. We now use Taylor series to approximate the functional form of the equitherms close to the maximum. By expanding each side of (14) in a Taylor series about the point $(z^*, 0)$, we obtain a relationship of the form

$$\frac{T_{\text{crit}}}{A} \left[R(0) + R'(0)r + \frac{R''(0)}{2!} r^2 + \cdots \right]$$
$$= F(z^*) + F'(z^*)(z - z^*) + \frac{F''(z^*)}{2!}(z - z^*)^2 + \cdots. \qquad (16)$$

a) Compute all the indicated derivatives and show that, by neglecting higher order terms, (16) implies that

$$\frac{T_{\text{crit}}}{A}(1 + \sigma r^2) \approx \frac{T_{\max}}{A}\left[1 - \frac{\gamma \mu^2}{2}(z - z^*)^2\right]. \qquad (17)$$

b) Show that this equation can be put into the standard form of an equation for an ellipse (in the zr-plane), i.e., it describes an ellipsoid of revolution in (r, z, θ) coordinates. You should find that the semimajor and minor axes of the ellipse are given by $2a$ and $2b$, where

$$a^2 = \frac{T_{\max} - T_{\text{crit}}}{\sigma^2 T_{\text{crit}}},$$
$$b^2 = \frac{2(T_{\max} - T_{\text{crit}})}{\gamma T_{\max} \mu^2}. \qquad (18)$$

This approximation verifies our guess that some equitherms are close to ellipsoidal.

12. The volume of an ellipsoid having dimensions a, b, and c (see **Figure 11**) is $\frac{4}{3}\pi abc$. What is the volume of the ellipsoid of revolution that we discovered in **Exercise 11**?

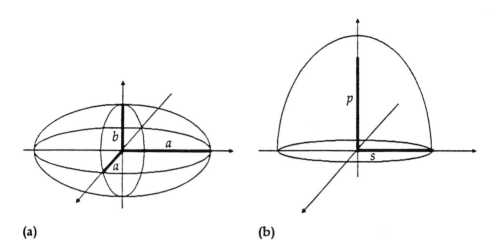

(a)   (b)

Figure 11. To estimate heated volume, we replace the cumbersome equitherms shown in Figure 10 (and described by (13)) by approximating surfaces, such as ellipsoids and paraboloids. The volumes enclosed by such simple surfaces are much easier to determine; see **Exercises 12** and **13**. [Adapted from Edelstein-Keshet et al. [1989]].

7.2 Some Equitherms Are (Almost) Parabolic

We next consider what happens when z is large in (13). Since, by assumption, $\omega^{1/2} > \mu$, the quantity $\exp(\mu - \omega^{1/2})z$ rapidly falls off to zero as z gets large. Thus, for large z, we have

$$r^2(z) \approx \frac{1}{\sigma^2}(C - \mu z), \tag{19}$$

where $C = \ln(A/T_{\text{crit}})$ is a constant, as before. This approximation establishes the fact that equitherms are parabolic at large depths.

Exercise

13. Convert (19) to the standard form for the equation of a paraboloid in cylindrical coordinates:

$$r^2 = s^2\left(1 - \frac{z}{p}\right).$$

What are s and p? Find the volume enclosed by the paraboloid surface of revolution (19) and the plane $z = 0$.

7.3 Estimating Heated Volumes

Our stated goal is to estimate V_T, the volume of tissue heated to temperatures above T_{crit}. To achieve this, we are faced with several possibilities:

- struggle with cumbersome integration of volumes inside the surfaces of revolution (**Exercise 9**);
- integrate numerically, with the help of a computer; or
- use simple shortcuts, replacing the true equitherms by approximating ellipsoids and paraboloids.

One advantage of the third alternative is that no integration is required: simple formulas for ellipsoidal and paraboloid volumes await us (**Exercises 12** and **13**), once we have a way to ascertain the chief dimensions of the surface, namely, a, b for the ellipsoid, and p, s for the paraboloid.

A further advantage of this third alternative is that it suggests a strategy for sampling tissue temperatures which will easily yield an estimate for the above quantities, as we will see in **Exercises 14** and **15**. In those exercises, we show that, given a critical temperature T_{crit}, the problem of estimating the volume of the region whose temperature exceeds T_{crit} reduces to estimating the lengths of the axes of an ellipsoidal or parabolic surface. We find that two measurement probes inserted parallel to each other at a known separation distance (with one of them along the axis of symmetry) give easily interpretable temperature data, that is, data that can be used in an elementary calculation of volume, easily implemented in a clinical situation.

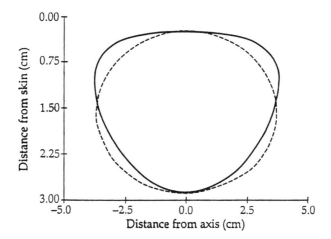

Figure 12. A comparison of the shape of one of the "true" equitherms (solid curve) (from **Figure 10**) with its ellipsoidal approximation (dotted curve). Note that the two share the same length and width but are otherwise not the same.

Approximations of somewhat irregular surfaces by ellipsoids and paraboloids help. But this is, after all, a shortcut that bears scrutiny. How well do such approximations describe the equitherms in an intermediate range, not close enough to z^* to be ellipsoidal yet not deep enough to be parabolic? **Figure 12** shows how true and ellipsoidal equitherms compare.

By construction, the two surfaces agree on the z-axis, although their shapes otherwise are not similar (since the distance away from the center z^* is large). Numerical integration techniques can be used to compare the volumes of the true and the approximating surfaces. Rather surprisingly, even for those equitherms whose shapes are not ellipsoidal—in other words, far from the temperature maximum—there is good agreement of the predicted volume over a wide range of temperatures. Apparently, places for which the true equitherm is "fatter" compensate for places at which the ellipsoid is "fatter" (see **Figure 13**).

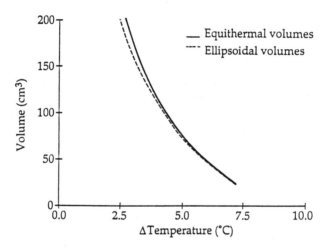

Figure 13. A comparison over a range of temperatures of the volumes of "true" equitherms with volumes of their ellipsoidal surrogates reveals surprisingly good agreement.

Exercises

14. **a)** Temperatures have been measured along a simple probe, inserted into the tissue along the z-axis at $r = 0$. **Figure 14** illustrates the temperature T as a function of the depth into the tissue. Use this graph to estimate b, the semimajor axis of the ellipsoid approximating the equitherm $T = 7.5°C$ above baseline. Can we get any information about the semi-major axis from this data?

   b) It is impractical to measure temperatures along the major axis of the ellipsoidal equitherm, since this direction is not easily accessible from

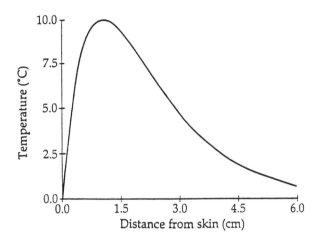

Figure 14. Temperature above baseline as measured along probe 1 in **Figure 15**. This curve was obtained by setting $r = 0$ in **(12)**. [Modified from Edelstein-Keshet et al. [1989]].

the skin surface. Instead, consider a second measurement probe parallel to the first and a distance $x_0 = 1.5$ cm from it (see **Figure 15**). Suppose $T = T_{\text{crit}}$ occurs at lengths ℓ_1 and ℓ_2 along the probe. Let $d = (\ell_2 - \ell_1)/2$, and use the equation of an ellipse to solve for a^2. Show that

$$a^2 = \frac{x_0^2}{1+u^2}, \quad \text{where} \quad u = \frac{d}{b}.$$

Use the above information and the formula for the volume of an ellipsoid to estimate the volume of the equitherm $T(r, z) = T_{\text{crit}}$. Compare the results obtained by using the above *two* probes to results that might be obtained from one or more probes placed at some skewed direction.

15. Suggest a measurement configuration for the parabolic-like equitherms which would achieve a similar result.

8. Epilogue

In this Module, we have used minimal reasonable assumptions to extrapolate from temperature distributions along line segments (the temperature probes) to temperature distributions inside a whole region representing a tumor. We showed that such assumptions could be borrowed either from geometric considerations about tumor shape (Section 6) or from specified properties of the heating pattern (Section 7). Instead of dwelling on the complete temperature distribution, we considered only a limited "descriptor" of the temperature distribution, namely the volume whose temperature exceeds

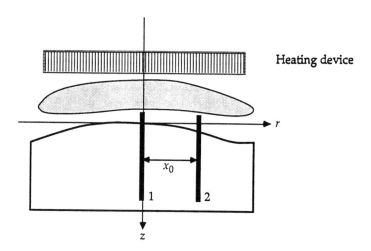

Figure 15. A configuration in which two probes are used. One has been inserted along the z-axis, the other at a distance x_0 from it.

T_{crit}. Finally, we exploited numerous shortcuts and simplifications to avoid having to actually perform unpleasant volume integrals. In some cases, we changed coordinates, while in others we substituted for the actual shape of the region a surrogate whose volume is easily determined.

Realistic clinical settings are as yet confounded by problems that limit the applicability of such simple approximations. As we indicated in the introduction, the variation of tissue properties, of blood-flow patterns, and of heating units, as well as errors of inaccurate probe insertion, are still among the chief problems faced in the clinical situation. Furthermore, whether the quantity V_T or V_T/V is truly descriptive of therapeutic treatment value is an issue that must still be resolved by experimental and statistical methods.

Yet, despite such limitations, the simple methods of calculus lead to a number of payoffs:

- We can dispel common clinical misconceptions (see **Exercise 1**).

- We can point to strategic locations for placing measurement probes (for example, along the axis of symmetry of the heating device, a direction that has not commonly been considered).

- We find that there are simple ways of extracting maximal information from a minimal data set (at least, in cases not too deviant from the ideal). Such methods are yet one more (albeit small) tool in a medical arsenal, much of which remains to be developed to combat the misery and suffering caused by diseases such as cancer.

9. Solutions to the Exercises

1. This conclusion is generally incorrect, since the tumor is three-dimensional. If the fraction of tumor radius heated successfully is f, then the fraction of the volume heated successfully would be proportional to f^3. For a spherical tumor of radius R, the volume corresponding to $R/3$ is $(R/3)^3 = 0.155R^3$.

2. Setting $T_{\text{crit}} = T(x, y, z)$ and rearranging terms leads to equations that describe a one-parameter family of surfaces, with $\beta \equiv T_{\text{max}} - T_{\text{crit}}$ the parameter.
    a) $\beta \equiv (T_{\text{max}} - T_{\text{crit}}) = \frac{x^2}{a^2} + \frac{y^2}{a^2} + \frac{z^2}{b^2}$ (concentric ellipsoids)
    b) $x^2 + y^2 + z^2 = \beta$ (concentric spheres)
    c) $\rho = \beta$ (concentric spheres in spherical coordinates)
    d) $\rho = \beta(1 + 0.2 \sin 8\theta \sin \phi)$ (a sphere with wrinkles (see **Figure 16**))
    e) $\rho = \beta(1 + 0.2 \sin 8\theta \sin 4\phi)$ (a sphere with bumps (see **Figure 16**))

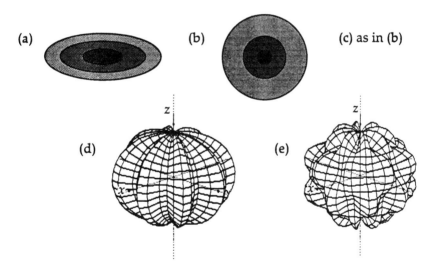

Figure 16. Solutions for Exercise 2.

3. $\alpha = \rho/R$, $\quad T(\alpha) = T_{\text{max}} - \dfrac{(T_{\text{max}} - T_{\text{edge}})}{R} R\alpha = T_{\text{max}} - m\rho$.

4. Eq. (2) implies (4). From (3) we get
$$T(\alpha) - T_{\text{max}} = -(T_{\text{max}} - T_{\text{edge}})\alpha_T,$$
implying (5).

5. Approximately 1.25 cm of the tumor radius has attained temperatures above 45°C. The fractional volume heated above 45°C is thus $(1.25/2.00)^3 = 0.244$: Only about one-fourth of the tumor volume has been successfully treated.

6. $T(\rho, \theta, \phi) = T = $ constant implies that

$$f\left(\frac{\rho}{R(\theta, \phi)}\right) = T,$$
$$f^{-1}(T) = \left(\frac{\rho}{R(\theta, \phi)}\right),$$
$$\rho = f^{-1}(T) \, R(\theta, \phi) = \alpha_T \, R(\theta, \phi),$$

where α_T is given by **(10)**.

7. The fraction of the tumor whose temperature exceeds T is

$$\frac{V_T}{V} = \frac{1}{V} \int_{\theta=0}^{2\pi} \int_{\theta=0}^{\pi} \int_{\rho=0}^{\alpha_T R(\theta,\phi)} dV,$$

where $dV = \rho^2 \sin\phi \, d\rho \, d\phi \, d\theta$ is the element of volume in spherical coordinates. Since f does not appear explicitly here, this integral can be evaluated to yield the desired result.

8. We assume, as in **Exercise 5**, that the critical temperature is 45°C and that the tumor has radius 2.0 cm. The temperature distribution in **Figure 8** reveals that an inner core of approximately 0.6 cm has been heated to temperatures exceeding 45°C. Thus,

$$\alpha_{45°} = f^{-1}(T) = f^{-1}(45°) = 0.6,$$
$$V_T/V = (0.6/2.0)^3 = 0.027.$$

Only about 3% of the tumor has been successfully treated.

9. a) Set $T(r, z) = T_{\text{crit}}$ and rearrange the equation to get

$$\exp(\sigma r)^2 = \left(\frac{A}{T_{\text{crit}}}\right) \exp(-\mu z) \left[1 - \exp\left(\mu - \omega^{1/2}\right) z\right].$$

Take the natural log of both sides to get

$$(\sigma r)^2 = \ln\left(\frac{A}{T_{\text{crit}}}\right) - \mu z + \ln\left[1 - \exp\left(\mu - \omega^{1/2}\right) z\right].$$

Divide by σ^2 to get **(13)**.

b) By the disk method, the volume of a surface of revolution is

$$V = \pi \int_p^q r^2(z) dz.$$

In order to compute the volume enclosed by an equitherm such as (13), we would need to integrate

$$\int C - \mu z + \ln\left[1 - \exp\left(\mu - w^{1/2}\right) z\right] dz.$$

The logarithm term makes this integration impractical. Further, we would have to determine values for the endpoints p and q, values of z satisfying $r^2(z) = 0$. Solving for them leads to a transcendental equation in z, which is a further difficulty.

10. a) The temperature distribution is given by (12), which can be written

$$T(r, z) = AF(z)/R(r).$$

A local maximum occurs when $\nabla T = \left(\frac{\partial T}{\partial r}, \frac{\partial T}{\partial z}\right) = 0$. Setting $F'(z) = 0$ leads to

$$0 = -\mu e^{-\mu z} + w^{1/2} e^{-w^{1/2} z}.$$

Rearranging twice and then taking natural logarithms of both sides gives successively

$$\begin{aligned}
\mu e^{-\mu z}) &= w^{1/2} e^{-w^{1/2} z}, \\
\mu/w^{1/2} &= \exp\left(\mu z - w^{1/2} z\right), \\
\ln\left(\mu/w^{1/2}\right) &= z(\mu - w^{1/2}),
\end{aligned}$$

which implies (15).

b)
$$\begin{aligned}
T_{max} &= T(z^*) \\
&= A e^{-(\sigma r)^2} \left(e^{\mu z^*} - e^{-w^{1/2} z^*}\right) \\
&= A e^{-\mu z^*} \left(1 - e^{(\mu - w^{1/2}) z^*}\right) \quad \text{(since } r = 0\text{)} \\
&= A e^{-\mu z^*} \left(1 - e^{-\ln \gamma}\right) \\
&= A e^{-\mu z^*} \left(1 - \frac{1}{\gamma}\right).
\end{aligned}$$

11. a) The Taylor series is computed about $z = z^*$, $r = 0$. The derivatives, evaluated at $(z^*, 0)$, are:

$$\begin{aligned}
F(z^*) &= T_{max}/A, \\
F'(z^*) &= 0 \quad \text{(by definition of } z^*\text{)}, \\
F''(z^*) &= \mu^2 e^{-\mu z^*} - w e^{-w^{1/2} z^*} = -\gamma \mu^2 (T_{max}/A); \\
R(0) &= 1, \\
R'(0) &= 0 \quad \text{(by definition)}, \\
R'' &= 2\sigma^2.
\end{aligned}$$

b) Eq. (17) implies that

$$1 + \sigma^2 r^2 \approx \frac{T_{max}}{T_{crit}} - \frac{\gamma \mu^2}{2} \frac{T_{max}}{T_{crit}} (z - z^*)^2,$$

$$\frac{r^2}{1/\sigma^2} + \frac{(z-z^*)}{2T_{\text{crit}}/\gamma\mu^2 T_{\text{max}}} = \frac{T_{\text{max}}}{T_{\text{crit}}} - 1.$$

Dividing both sides by $(T_{\text{max}} - T_{\text{crit}})/T_{\text{crit}}$ leads to

$$\frac{r^2}{a^2} + \frac{(z-z^*)^2}{b^2} = 1,$$

where a^2 and b^2 are given by (18).

12. $V = \frac{4\pi}{3}a^2 b$, since the z-axis is an axis of symmetry, so that the third dimension of the ellipsoid is a.

13. $r^2 = \frac{c}{\sigma^2}\left(1 - \frac{\mu}{z}\right)$, $s = \sqrt{C}/\sigma$, $p = C/\mu$.

14. a) The distance along the catheter between the two places at which $T = 7.5°C$ above normal (baseline) temperature is approximately 1.5 cm. Thus, $b \approx \frac{1.5}{2} = 0.75$ cm, which would be the semi-axis of the ellipsoid in the z-direction. We can get no information about a, the semi-axis of the ellipsoid in the radial direction, from these data.

b) The equation of the elliptical cross-section is

$$\frac{r^2}{a^2} + \frac{(z-z^*)^2}{b^2} = 1.$$

The r-coordinate of the second probe is x_0, and the points on the catheter at which T_{crit} is attained have z coordinates $z^* + d$ and $z^* - d$. Thus, the equation becomes

$$\frac{x_0^2}{a^2} + \frac{d^2}{b^2} = 1.$$

We can use this equation to solve for a^2, obtaining the result given in the exercise. The volume of the equitherm is then

$$V = \frac{4\pi}{3}a^2 b = \frac{4\pi}{3}\frac{x_0^2 b^3}{b^2 - d^2}.$$

Placing the second catheter in parallel with the first makes it easy to apply the equation of an ellipse to find the length of the semi-axis a. If the catheter is skewed, it is more difficult to determine the coordinates of the points at which the temperature T_{crit} is attained. We can use trigonometry to calculate the coordinates, if we know the angle made by the catheter relative to the z-axis; but doing so may introduce further error and uncertainty into the calculations.

15. An analogous configuration for the parabolic equitherms would be to have one catheter along the axis of the paraboloid and another at some distance *d* away from this axis and parallel to it.

References

Dewhirst, M.W., J.M. Winget, L. Edelstein-Keshet, J. Sylvester, M. Engler, D.E. Thrall, R.L. Page, and J.R. Oleson. 1987. Clinical application of thermal isoeffect dose. *International Journal of Hyperthermia* 3: 307–318.

Edelstein-Keshet, L., M.W. Dewhirst, J.R. Oleson, and T.V. Samulski. 1989. Characterization of tumor temperature distributions in hyperthermia based on assumed mathematical forms. *International Journal of Hyperthermia* 5: 757–777.

Hahn, M. 1982. *Hyperthermia and Cancer*. New York: Plenum Press.

Samulski, T.V., R.S. Cox, B.E. Lyons, and P. Fessenden. 1988. Heat transfer and blood flow during hyperthermia in normal canine brain: II Mathematical model. *International Journal of Hyperthermia* 5: 249–263.

Storm, F. 1983. *Hyperthermia in Cancer Therapy*. Boston, MA: G.K. Hall Medical Publishers.

Streffer, C., ed. 1987. *Hyperthermia and the Therapy of Malignant Tumors*. New York: Springer-Verlag.

Watmough, D.J., and W.M. Ross., eds. 1986. *Hyperthermia*. Glasgow: Blackie.

Acknowledgments

The author wishes to express sincere gratitude to Drs. D.W. Dewhirst, J.R. Oleson, and T.V. Samulski for the opportunity to take part in their unique and stimulating team effort. For memorable modeling discussions, for comments and insights, for permission to publish this material in a format suitable to mathematics students, and for several original figures, I am particularly indebted. Joint work on this problem was supported by NIH–NCI grant 5P01CA 42745–03 and NSF grant DNS–86–01644. The author is currently supported by Canadian NSERC operating grant OG PIN 021.

About the Author

Leah Edelstein-Keshet first became involved in mathematical modeling of biological phenomena as a Master's student at Dalhousie University (Halifax, Nova Scotia, Canada) and later as a Ph.D. student in applied mathematics at the Weizmann Institute of Science (Rehovot, Israel). She has since worked at Brown and Duke Universities, often in collaboration with biologists or medical doctors. Work on models for hyperthermia evolved through regular meetings of the "Math Group" in the Radiation Oncology Department of the Duke Medical Center, Durham, NC. The author's current interests include developmental biology, population dynamics, and pattern formation. She is the author of *Mathematical Models in Biology* (Random House, 1988). At the University of British Columbia, she continues to teach, carry out research, and participate in the development of a graduate studies program in mathematical biology.

UMAP

Modules in Undergraduate Mathematics and its Applications

Published in cooperation with the Society for Industrial and Applied Mathematics, the Mathematical Association of America, the National Council of Teachers of Mathematics, the American Mathematical Association of Two-Year Colleges, The Institute of Management Sciences, and the American Statistical Association.

Module 716

Newton's Method and Fractal Patterns

Philip D. Straffin, Jr.

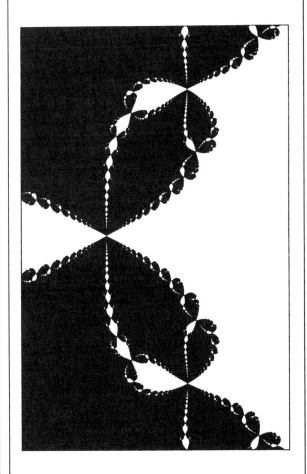

Applications of Calculus to Fractals and Chaotic Dynamics

COMAP, Inc., Suite 210, 57 Bedford Street, Lexington, MA 02173 (617) 862–7878

| INTERMODULAR DESCRIPTION SHEET: | UMAP Unit 716 |

TITLE: Newton's Method and Fractal Patterns

AUTHOR: Philip D. Straffin, Jr.
Department of Mathematics and Computer Science
Beloit College
700 College St.
Beloit, WI 53511–5595

MATHEMATICAL FIELD: Calculus

APPLICATION FIELD: Fractals, chaotic dynamics

TARGET AUDIENCE: First-semester calculus students.

ABSTRACT: Newton's method is an efficient iterative procedure for approximating zeros of a differentiable function. When the function has several zeros, which zero will be found depends on where we start the Newton iteration. The resulting behavior of "Newton's method in the large" is intricate even for real polynomials, and in the complex plane it generates beautiful fractal patterns.

PREREQUISITES: Derivatives of polynomials and rational functions, definition of the derivative, the derivative as a linear approximation. A computer with appropriate software, or at least a calculator, will be helpful for doing the exercises.

©Copyright 1991, 1992 by COMAP, Inc. All rights reserved.

COMAP, Inc., Suite 210, 57 Bedford Street, Lexington, MA 02173
(800) 77–COMAP = (800) 772–6627,   (617) 862–7878

Newton's Method and Fractal Patterns

Philip D. Straffin, Jr.
Dept. of Mathematics and Computer Science
Beloit College
700 College St.
Beloit, WI 53511-5595

Table of Contents

1. How Does Newton's Method Behave in the Large? 1
2. Newton's Method 1
3. Newton's Method in the Large: An Example on the Real Line . 5
4. Newton's Method in the Complex Plane 9
5. Further Directions 15
6. Solutions to the Exercises 16
    References 18
    Acknowledgments 18
    About the Author 18

MODULES AND MONOGRAPHS IN UNDERGRADUATE
MATHEMATICS AND ITS APPLICATIONS (UMAP) PROJECT

The goal of UMAP is to develop, through a community of users and developers, a system of instructional modules in undergraduate mathematics and its applications, to supplement existing courses and from which complete courses may be built.

The Project was initially funded by a grant from the National Science Foundation and has been guided by a National Advisory Board of mathematicians, scientists, and educators. UMAP is now supported by the Consortium for Mathematics and Its Applications (COMAP), Inc., a non-profit corporation engaged in research and development in mathematics education.

COMAP Staff

Paul J. Campbell	Editor
Solomon Garfunkel	Executive Director, COMAP
Laurie W. Aragón	Development Director
Philip A. McGaw	Production Manager
Roland Cheyney	Project Manager
Laurie M. Holbrook	Copy Editor
Dale Horn	Design Assistant
Rob Altomonte	Distribution Coordinator
Sharon McNulty	Executive Assistant

1. How Does Newton's Method Behave in the Large?

Some of the oldest problems in mathematics involve finding solutions to equations of the form $f(x) = 0$. Such solutions are called *zeros* of f, or sometimes *roots* of f. For a polynomial of degree one or two, general methods of finding zeros were known before 2000 B.C. In the 16th century, Italian mathematicians dal Ferro, Tartaglia, Cardano, and Ferrari developed methods to find exact zeros of polynomials of degrees three and four. However, since the work of Abel in 1826, we have known that there is no general method for solving exactly polynomial equations of degree greater than or equal to five.

In cases where we cannot solve $f(x) = 0$ exactly, we need an efficient method of approximating solutions to any desired degree of accuracy. (Indeed, such a method is valuable even when we can solve the equation exactly; for if the solution involves, say, cube roots, how can we calculate them efficiently?) Isaac Newton found just such a method, based on his newly developed differential calculus, in 1669. In an improved form due to Joseph Raphson in 1690, this method is now taught in beginning calculus courses as *Newton's method*. We will present it in the next section. It involves choosing an initial guess x_0, and finding iteratively a sequence of numbers $x_1, x_2, x_3, \ldots$ that converge to a solution.

When the function $f(x)$ has several zeros, *which zero* Newton's method will find depends on the initial guess x_0. The pattern of which initial guesses lead to which zeros—the behavior of Newton's method "in the large"—turns out to be surprisingly complicated and interesting even for polynomials. When we generalize slightly and apply Newton's method to polynomials $f(z)$ as functions of a complex number z (which we picture as a point in the complex plane), the behavior of Newton's method in the large produces pictures that are infinitely complicated and astonishingly beautiful. A sample is shown at the end of this Module. Understanding the mathematics behind these pictures is a subject of current research, and the mathematics has strong ties to the study of general chaotic systems. The goal of this Module is to guide you along this surprisingly short road from beginning calculus to a research frontier of mathematics.

2. Newton's Method

Suppose we have a differentiable function $f(x)$ for which we wish to find a zero. We start with an initial point x_0, and we determine a new point x_1 by beginning at the point $(x_0, f(x_0))$ on the graph of f and following the tangent line from this point to where it intersects the x-axis (see **Figure 1**). Since the

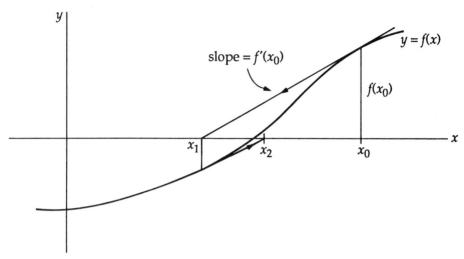

Figure 1. The geometry of Newton's method.

slope of the tangent line is $f'(x_0)$, we have

$$\frac{f(x_0)}{x_0 - x_1} = f'(x_0),$$

so that

$$x_1 = x_0 - \frac{f(x_0)}{f'(x_0)}.$$

We then use x_1 as the starting point for the next iteration of this procedure, to get x_2. Thus we generate a sequence of points x_n by the rule

$$x_{n+1} = x_n - \frac{f(x_n)}{f'(x_n)}.$$

If we choose the initial point x_0 close to the zero x_* that we are trying to locate, the x_n's will converge to x_* quite rapidly.

For the purposes of this Module, it will be convenient to have a more compact notation for the Newton iteration. Given a differentiable function f, define the *Newton function for f* by

$$N(x) = x - \frac{f(x)}{f'(x)}.$$

Then $x_1 = N(x_0)$, $x_2 = N(x_1) = N(N(x_0)) = N^2(x_0)$, and, in general,

$$x_n = N^n(x_0),$$

where the notation N^n means "N applied n times."

As an example of Newton's method, let us approximate the solution to $x^3 - x - 1 = 0$. A quick graph of $f(x) = x^3 - x - 1$ shows that it has just one zero, between 1 and 2. The Newton function for f is

$$N(x) = x - \frac{x^3 - x - 1}{3x^2 - 1} = \frac{2x^3 + 1}{3x^2 - 1}.$$

Starting with $x_0 = 1$, the results of the Newton calculations are

$x_0 = $ 1
$x_1 = $ 1.5
$x_2 = $ 1.34...
$x_3 = $ 1.3252...
$x_4 = $ 1.3247181...
$x_5 = $ 1.324717957244789...
$x_6 = $ 1.32471795724474602596091...
$x_7 = $ 1.32471795724474602596091....

Since x_6 and x_7 agree to the shown accuracy, we conclude that we have found the solution to this accuracy. Notice how quickly the sequence converged once we got close to the zero. The number of correct decimal places roughly *doubled* with each iteration: x_2 was correct to 1 decimal place, x_3 to 2 places, x_4 to 5 places, x_5 to 13 places, and x_6 to at least 23 places. This kind of convergence is called *quadratic convergence* by numerical analysts, and it is characteristic of Newton's method.

We can understand why Newton's method usually works so well by using some calculus on the Newton function. First of all, from the definition

$$N(x) = x - f(x)/f'(x),$$

we see that x_* is a zero of f if and only if $N(x_*) = x_*$, i.e., x_* is a *fixed point* of N. Next, consider the distance from x_n to the fixed point x_*:

$$x_{n+1} - x_* = N(x_n) - N(x_*) = N'(x_*)(x_n - x_*) + \mathcal{O}\left((x_n - x_*)^2\right).$$

The second equality is just the linear approximation property of the derivative, with the notation $\mathcal{O}(h^2)$ meaning a quantity that approaches zero like h^2 when h gets small. Notice that if x_n is close to x_*, then x_{n+1} will be even closer if $|N'(x_*)| < 1$. In this case, x_* is called an *attracting fixed point* of N. So we should compute the derivative of N.

By the quotient rule,

$$N'(x) = 1 - \frac{f'(x)f'(x) - f(x)f''(x)}{[f'(x)]^2} = \frac{f(x)f''(x)}{[f'(x)]^2},$$

at least when $f'(x) \neq 0$. Thus, if x_* is a zero of f, we find that $N'(x_*) = 0$. This is certainly less than 1. In fact, it means that

$$x_{n+1} - x_* = \mathcal{O}\left((x_n - x_*)^2\right),$$

which is exactly the quadratic convergence property for Newton's method. Since $N'(x_*) = 0$ gives quadratic convergence, we say that x_* is a *super-attracting fixed point* of N. What we have said is important enough to summarize in a theorem. Recall that a *critical point* of f is a value x such that $f'(x) = 0$.

Theorem. *Suppose f is a differentiable function. Then a number x_* is a zero of f if and only if it is a fixed point of the Newton function N. If x_* is a zero of f that is not also a critical point of f, then x_* is a super-attracting fixed point of N.*

It is only fair to point out that Newton's method does not always work so beautifully. For one thing, our argument for quadratic convergence required that $f'(x_*) \neq 0$. If $f'(x_*) = 0$, convergence to x_* is much slower (see **Exercise 3**).

Even worse, for some choices of the initial guess x_0, Newton's method may not converge at all. In the example above, $f(x)$ has critical points at $x = \pm 1/\sqrt{3}$. If we should choose x_0 to be one of those critical points, $N(x_0)$ will be undefined and Newton's method will fail. In fact, it will fail if any x_n is a critical point. Finally, there are cases in which the entire sequence of x_n's is defined but does not converge to a zero of f. The simplest case is when the sequence of x_n's becomes *periodic*. For instance, we will see a case shortly in which the sequence is periodic with period two: $x_0 = x_2 = x_4 = \ldots$. It is also possible for the Newton sequence to jump about chaotically, never settling down to any regular behavior. **Exercises 4** through **6** ask you to think about some of these difficulties. The moral for practical users of Newton's method is clear. *Choose your initial guess wisely, close to where you know there is a zero!*

Exercises

1. Explain why, if $x_{n+1} - x_* \approx (x_n - x_*)^2$, the number of correct decimal places in x_n should approximately double with each iteration.

2. For the following functions, calculate the Newton function and iterate it starting at the given initial point. Look for quadratic convergence to a zero of f.

    a) $f(x) = x^2 - 2$; $x_0 = 1$
    b) $f(x) = x^3 - 2x$; $x_0 = 1$
    c) $f(x) = x^3 - 2x$; $x_0 = 0.7$
    d) $f(x) = x^3 - 2x$; $x_0 = 0.5$

3. The function $f(x) = x^2$ has a unique zero at $x = 0$, which is also a critical point.

    a) Find the Newton function N for f. Show that 0 is an attracting fixed point of N, but not a super-attracting fixed point.

b) With initial point $x_0 = 0.04$, correct to one decimal place, we would expect, if quadratic convergence held, that x_4 would be correct to 16 decimal places. How many iterates must we really do to approximate the zero to 16 decimal places? (Hint: Think rather than just iterate!)

4. Use the "follow the tangent line" description of Newton's method to explain geometrically why the method fails when $f'(x_n) = 0$ for some n.

5. Try to sketch the graph of a function f, an initial point x_0, and tangent lines giving x_1 and x_2 with $x_2 = x_0$, so that the Newton sequence is periodic with period two.

6. Calculate the Newton function for $f(x) = x^2 + 1$. Iterate it starting at $x_0 = 0.5$ for as long as you have patience. Do you see any patterns or tendency to converge in the Newton sequence? Why should we expect trouble here?

3. Newton's Method in the Large: An Example on the Real Line

In **Exercises 2b–d**, you saw that when $f(x)$ has several zeros, different initial points x_0 can lead to different zeros. The pattern of which initial points lead to which zeros is not simple. **Exercise 2c**, for example, shows that Newton's method need not converge to the zero that is closest to x_0. In this section, we will study the pattern of convergence to zeros for a slightly simpler polynomial, $f(x) = x^3 - x$. This function clearly has zeros at $x = -1, 0, 1$. *Finding* the zeros is no problem! Rather, we will be concerned with *which initial points lead to which zeros*.

Definition. *If x_* is a zero of f, the **basin of attraction** of x_* is the set of all numbers x_0 such that Newton's method starting at x_0 converges to x_*. In symbols,*
$$B(x_*) = \{x_0 | \, x_n = N^n(x_0) \text{ converges to } x_*\}.$$

We want to calculate the basins of attraction of -1, 0, and 1. Since these points are all attracting fixed points (in fact, super-attracting fixed points) of N, we know that some open interval around each one is contained in its basin of attraction. The *largest* such interval is called the *local basin of attraction*. We will start by finding the local basins.

From the graph of f in **Figure 2**, it is clear that if $x_0 \geq 1$, x_n will converge to 1. In our notation, $[1, \infty) \subset B(1)$. Moreover, if x_0 is between the critical point $1/\sqrt{3}$ and 1, it will be true that $x_1 > 1$, so that x_n will still converge to 1. Hence $(1/\sqrt{3}, \infty) \subset B(1)$. If $x_0 = 1/\sqrt{3}$, Newton's method fails, so this is the largest open interval about 1 which is contained in $B(1)$: it is the local basin of attraction for 1. Similarly, $(-\infty, -1/\sqrt{3}) \subset B(-1)$ is the local basin of attraction of -1.

Now consider the zero of f at $x = 0$. A little experimentation, or a careful look at the graph in **Figure 2**, shows that points close to 0 oscillate around 0: if $x > 0$, then $N(x) < 0$, and vice versa. This suggests that we might look for a point of period 2 for N, i.e., a point x for which $N^2(x) = x$. To do this, we first calculate

$$N(x) = x - \frac{x^3 - x}{3x^2 - 1} = \frac{2x^3}{3x^2 - 1}.$$

Notice by symmetry that we will have $N^2(x) = x$ if we have $N(x) = -x$. (The symmetry simplifies matters greatly. It happens because f, and hence N, is an odd function: $N(-x) = -N(x)$.) Solving for x, we get

$$-x = \frac{2x^3}{3x^2 - 1}, \qquad 5x^3 - x = 0, \qquad x = 0, \ \pm 1/\sqrt{5}.$$

Hence $\pm 1/\sqrt{5}$ are points of period two for N. Furthermore, you can check (**Exercise 7**) that if $|x| < 1/\sqrt{5}$, then $|N(x)| < |x|$. We have found that the local basin for 0 is $(-1/\sqrt{5}, 1/\sqrt{5})$. The local basins of attraction are shown in **Figure 2**.

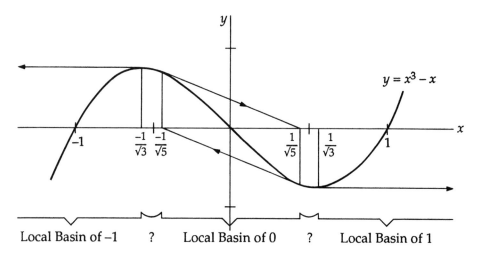

Figure 2. Local basins for Newton's method.

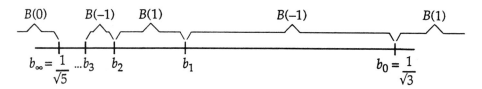

Figure 3. Basin structure in $\left(\frac{1}{\sqrt{5}}, \frac{1}{\sqrt{3}}\right)$ (distances are not to scale).

The interesting behavior of Newton's method for this example occurs in the interval $(1/\sqrt{5}, 1/\sqrt{3})$ and the symmetric negative interval. Let's look at these intervals. It should be clear from **Figure 3** that if we move x slightly to the left of $1/\sqrt{3}$, then $N(x)$ will be large and negative, so that x will be in $B(-1)$. As we continue to decrease x, it will stay in $B(-1)$ until $N(x) = -1/\sqrt{3}$. We can solve the equation

$$N(x) = \frac{2x^3}{3x^2 - 1} = -\frac{1}{\sqrt{3}} = -0.577350$$

to find $x \approx 0.465601$. Thus the interval $(0.465601, 0.577350)$ is in $B(-1)$, and by symmetry the interval $(-0.577350, -0.465601)$ is in $B(1)$.

As we decrease x past 0.465601, $N(x)$ moves past -0.577350 into $B(1)$, so x is in $B(1)$. It stays in $B(1)$ until

$$N(x) = \frac{2x^3}{3x^2 - 1} = -0.465601,$$

which happens at $x \approx 0.450202$. In general, we find a sequence of numbers $b_0 = 1/\sqrt{3} > b_1 \approx 0.465601 > b_2 \approx 0.450202 > b_3 > \ldots$ such that

$$(b_i, b_{i-1}) \subset B(-1) \text{ when } i \text{ is odd,}$$

and

$$(b_i, b_{i-1}) \subset B(1) \text{ when } i \text{ is even.}$$

The numbers b_i can be determined by successively solving equations $N(b_i) = -b_{i-1}$.

The values of the first few b_i's are given in **Table 1**, along with the lengths of the intervals (b_i, b_{i-1}) and the ratios of lengths of successive intervals. Notice the interesting behavior. Each of $B(-1)$ and $B(1)$ consists of infinitely many intervals, whose lengths decrease approximately geometrically. An arbitrarily small movement of x_0 to the right of $1/\sqrt{5}$ will cause convergence to shift between 1 and -1 infinitely often.

Table 1.
Lengths of intervals and ratios of lengths of successive intervals.

i	b_i	$b_i - b_{i-1}$	$(b_i - b_{i-1})/(b_{i+1} - b_i)$
0	.577350		
1	.465601	.111749	7.26
2	.4502020	.015399	6.18
3	.4477096	.0024924	6.03
4	.4472962	.0004134	6.01
5	.44722736	.00006884	6.00
6	.44721589	.00001147	6.00
7	.44721398	.00000191	6.00
⋮	⋮	⋮	⋮
∞	.447213595		

Exercises

7. Check the claim that if $|x| < 1/\sqrt{5}$, then $|N(x)| < |x|$.

8. What happens to Newton's method for $f(x) = x^3 - x$ if we choose $x_0 = b_i$ for some i?

9. Verify that $b_1 \approx 0.465601$ by using Newton's method(!) to approximate the real solution to $2x^3 + 0.577350(3x^2) - 0.577350 = 0$. (See where this comes from?)

10. Let's see why the ratios of lengths of successive intervals in **Table 1** approach 6.
    a) Verify that $N'(1/\sqrt{5}) = -6$. From the definition of the derivative, this means that if x' and x'' are any two points close to $1/\sqrt{5}$, we will have
    $$\frac{N(x') - N(x'')}{x' - x''} \approx -6.$$
    b) Now use the fact that $N(b_i) = -b_{i-1}$, and the fact that b_i is close to $1/\sqrt{5}$ for large i, to show that for large i,
    $$\frac{b_i - b_{i-1}}{b_{i+1} - b_i} \approx 6.$$

11. What are the Newton's method basins of attraction for the roots of $x^2 - 2$? What about for the roots of $x^2 - 4x + 3$? Justify your answers.

4. Newton's Method in the Complex Plane

I think you will agree that the global behavior of Newton's method can be interesting on the line of real numbers. However, it is in the plane of complex numbers that we see the true intricacy and beauty of the patterns the method can generate.

Recall that a complex number z has the form $z = x + iy$, where x and y are real numbers and i is a symbol having the property that $i^2 = -1$. We call x the *real part*, and y the *imaginary part*, of z. We represent the complex number $z = x + iy$ as the point (x, y) on a coordinate plane that we call the *complex plane*. The x-axis is called the *real axis*; the y-axis is the *imaginary axis*. The *norm* of a complex number is the non-negative real number $|z| = \sqrt{x^2 + y^2}$. Geometrically, it is the distance from z to the origin $0 = 0 + 0i$.

Addition and subtraction of complex numbers is done componentwise, so that if $z = x + iy$ and $w = u + iv$, then

$$z + w = (x + u) + i(y + v),$$
$$z - w = (x - u) + i(y - v).$$

Note that $|z - w|$ is the geometric distance between z and w. Multiplication is done using the distributive laws and the property that $i^2 = -1$:

$$zw = (x + iy)(u + iv) = xu + i(xv + yu) + i^2(yv) = (xu - yv) + i(xv + yu).$$

To divide complex numbers, we use the standard method of "rationalizing the denominator":

$$\frac{z}{w} = \frac{x+iy}{u+iv} \cdot \frac{u-iv}{u-iv} = \frac{(xu+yv)+i(yu-xv)}{u^2+v^2} = \frac{xu+yv}{u^2+v^2} + i\left(\frac{yu-xv}{u^2+v^2}\right).$$

Derivatives of functions of a complex variable behave computationally just like derivatives of functions of a real variable. For example, if $f(z) = z^3 - z$, then $f'(z) = 3z^2 - 1$. Moreover, Newton's method generalizes directly to the complex plane: if $N(z) = z - f(z)/f'(z)$, and z_0 is a complex number, then the iterates $N^n(z_0)$ will in general converge quadratically, in norm, to a zero of $f(z)$.

For a first example of Newton's method in the complex plane, let's consider $f(z) = z^2 + 1$. The corresponding real function $f(x) = x^2 + 1$ has no real roots, and you saw in **Exercise 6** that Newton's method responded to this situation by jumping chaotically around on the real line. However, $z^2 + 1 = 0$ has two solutions, at $z = i$ and $z = -i$. If we choose z_0 on the real axis, the Newton iterates will do exactly as they did in **Exercise 6**, since on the real axis, complex arithmetic reduces to real arithmetic. However, if we choose z_0 off the real axis, Newton's method converges nicely. For example:

$z_0 =$	1	$+ 0.5i$	$z_0 =$	$0.5 \quad - i$
$z_1 =$	$+0.1000$	$+ 0.4500i$	$z_1 =$	$+0.0500 - 0.9000i$
$z_2 =$	-0.1853	$+ 1.2838i$	$z_2 =$	$-0.0058 - 1.0038i$
$z_3 =$	-0.0376	$+ 1.0234i$	$z_3 =$	$-i$
$z_4 =$	-0.0009	$+ 0.9996i$		
$z_5 =$	i			

If we apply Newton's method to $f(z) = z^3 - z$ with initial points off the real axis, we notice that global behavior can be delicate and not easily predictable. We have, for example, the values below, with the behavior shown in **Figure 4**:

$z_0 =$	$+0.60 \quad + 0.45i$	$+0.65 \quad + 0.45i$	$0.70 \quad + 0.45i$
$z_1 =$	$+0.4947 + 0.0222i$	$+0.5520 + 0.0301i$	$0.6067 + 0.0429i$
$z_2 =$	$-0.8207 - 0.3249i$	$-1.3527 - 2.1411i$	$1.7051 - 1.7393i$
$z_3 =$	$-0.7866 + 0.0200i$	$-0.9442 - 1.3527i$	$1.1967 - 1.0911i$
$z_4 =$	$-1.1320 - 0.0381i$	$-0.6910 - 0.7892i$	$0.8870 - 0.6224i$
$z_5 =$	$-1.0190 - 0.0103i$	$-0.5437 - 0.3489i$	$0.7267 - 0.2445i$
$z_6 =$	$-1 \quad - 0.0006i$	$-0.4224 + 0.1111i$	$0.7689 + 0.1833i$
$z_7 =$	-1	$+0.0823 - 0.2777i$	$0.8965 - 0.1785i$
$z_8 =$		$+0.0272 - 0.0291i$	$0.9371 + 0.0383i$
$z_9 =$		$+0.0001 + 0.0001i$	$1.0034 - 0.0088i$
$z_{10} =$		0	$0.9999 - 0.0001i$
$z_{11} =$			1

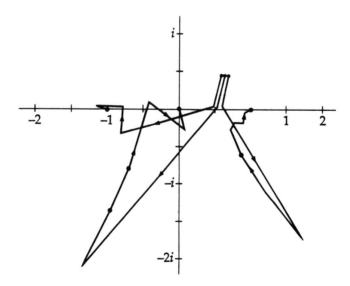

Figure 4. Newton's method for $z^3 - z$.

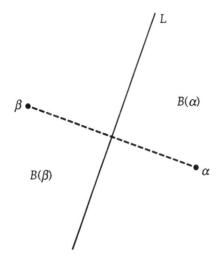

Figure 5. Newton basins for a quadratic polynomial.

Can we make some sense out of this complicated behavior and understand the global behavior of Newton's method in the complex plane? In particular, can we find the basins of attraction of the different zeros of $f(z)$, at least for simple functions f? This question was first considered by the English mathematician Arthur Cayley more than a hundred years ago. Cayley will become a familiar name to you if you study linear algebra, or graph theory as it is used in computer science. In the 1850s, Cayley first defined matrix multiplication and proved the fundamental theorems about matrices, as well as defining the concept of a "tree" and analyzing basic properties of trees.

Cayley [1879] realized that the Newton-Raphson method could be used to approximate roots of functions of a complex variable and posed the problem of determining the basins of attraction of the different roots. The problem, by the way, was posed in Vol. 2 of the *American Journal of Mathematics*, the first American professional journal in mathematics, which had just been founded at Johns Hopkins University by Cayley's friend J.J. Sylvester.

Cayley was able to solve the problem completely for quadratic polynomials. He proved the result that you might have guessed in the real case from **Exercise 11**.

Theorem (Cayley, 1879). *Let the complex quadratic polynomial $f(z) = az^2 + bz + c$ have zeros α and β in the complex plane. Let L be the perpendicular bisector of the line segment from α to β. Then, when Newton's method is applied to $f(z)$, the half-planes into which L divides the complex plane are exactly $B(\alpha)$ and $B(\beta)$, the basins of attraction to α and β.*

The basins are pictured in **Figure 5**. Another way to state the result is that Newton's method starting at z_0 will converge to α precisely when $|z_0 - \alpha| < |z_0 - \beta|$. On the bisector L, we now know that the Newton function

$N(z)$ is "chaotic" in the technical sense of the modern theory of chaos. This explains the result of **Exercise 6**: the real axis is the perpendicular bisector of the line segment between the two zeros i and $-i$ of $z^2 + 1$.

Our example on the real line shows that the situation for cubic polynomials must be more complicated: initial points will not always converge to the closest zero of f. Cayley considered cubic polynomials, but he wrote that in this case "it is anything but obvious what the division is, and the author has not succeeded in finding it." In fact, even the qualitative nature of the Newton basins for a cubic polynomial wasn't understood until the work of Fatou and Julia on "Julia sets" about 1918; and the first real pictures of these basins were drawn only in the 1980s, when powerful computer graphics became available.

Let's consider the cubic polynomial $f(z) = z^3 - z$. To picture the global behavior of Newton's method in this example, we ask a computer to color each point z_0 in the complex plane according to which zero of $z^3 - z$ Newton's method will converge to, if we start at z_0. **Figure 6** shows the result, with $B(-1)$ colored gray, $B(0)$ colored black, and $B(1)$ colored white. Notice how the pattern we found on the real line extends to an intricate fractal pattern in the complex plane. **Figure 7** shows a blow-up of the gray bulb near $z_0 = 0.5$. Each gray bulb has infinitely many white bulbs attached densely along its boundary, all of those white bulbs have infinitely many gray bulbs attached densely along their boundaries, and so on *ad infinitum*. No wonder Cayley had trouble picturing the shapes of the basins!

It turns out that at any point where two of the basins meet, the third basin does as well. **Figure 6** is one solution to the problem of dividing a plane into three regions such that all three regions share a common boundary. The infinitely loopy fractal set that is the common boundary is called a *Julia set*. The Newton function $N(z) = 2z^3/(3z^2 - 1)$ takes the Julia set onto itself, and does so chaotically. It also preserves each of the three colored regions, and transforms them in interesting ways. For instance, N takes the gray bulb around $z = 0.5$ onto the main gray region at the left. N takes three different bulbs onto the bulb around $z = 0.5$: the bulbs around $z = 0.6 + 0.45i$ (see **Figure 4**), $z = 0.6 - 0.45i$ and $z = -0.46$ (see **Figure 3** and **Table 1**).

While **Figure 6** would probably not win any beauty contests, Newton basins for other cubic and higher order polynomials can have quite beautiful structures. **Figure 8** is an example from Peitgen and Richter [1986]. It shows part of the basins for Newton's method applied to the cubic polynomial $z^3 - 1$. $B(1)$ is in white, and the other two basins are both colored black. The origin is at the center of the picture.

Newton's Method and Fractal Patterns 119

Figure 6. Newton basins for $z^3 - z$.

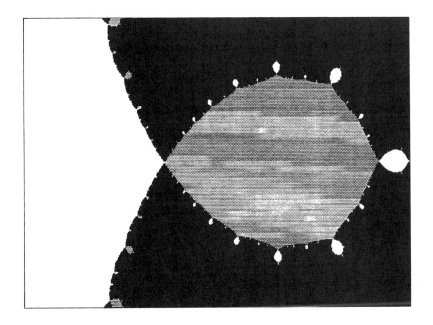

Figure 7. Detail of a bulb from **Figure 6**.

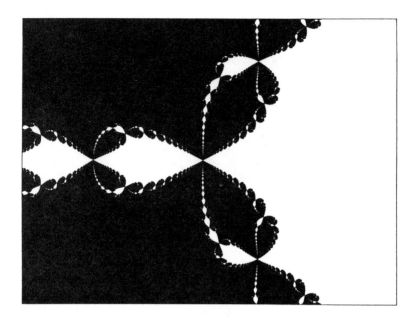

Figure 8. Newton basins for $z^3 - 1$. [From Peitgen and Richter [1986]].

Exercises

12. We started our discussion with quadratic polynomials. What about linear polynomials? For $f(z) = az + b$, compute $N(z)$. What is the basin of attraction of the zero $-b/a$?

13. If $f(z) = z^2 + 1$, calculate the first three Newton iterates starting from $z_0 = 0.5i$. Are they behaving as Cayley's Theorem says they should be?

14. If $f(z) = z^2 + 1$ and $z = x + iy$, find the real and imaginary parts of $N(z)$ as functions of x and y. (This is what you need to do if you are going to program a computer to carry out Newton's method, if your computer does not support complex arithmetic.)

15. Use the formula for the imaginary component in **Exercise 14** to show that $N(z)$ maps $B(i)$ (the upper half-plane) into itself.

16. If you succeeded well enough in **Exercise 14** to want to try something harder, program a computer to do Newton's method on $f(z) = z^3 - z - 1$. Find the three complex zeros of f.

17. Program a computer to do Newton's method on $f(z) = z^3 - 1$ and find three starting values z_0, all with *positive* real part, which converge to the three different zeros of f. (**Figure 8** can help you.)

5. Further Directions

Peitgen and Richter [1986, Chapter 6] and Peitgen et al. [1984] have a number of other pictures of Newton's method in the complex plane, and there is also a discussion in Gleick [1987]. Becker and Dörfler [1989, Chapter 4] have a number of do-it-yourself computer experiments involving Newton's method and other ways of generating fractals. Saari and Urenko [1984] survey some recent results about Newton's method on the real line. Strang [1991] shows a number of interesting examples of Newton's method, emphasizing chaotic behavior.

The global behavior of Newton's method in the complex plane is an area of current research in mathematics. For instance, in the examples we have seen, the initial points z_0 for which Newton's method fails to converge to any zero of $f(z)$, are clearly rare. They are points on the the boundary between different basins, and you would have to be very unlucky to choose one by chance. However, in 1983 Curry et al. [1983] discovered that there are many cubic polynomials for which another kind of failure is possible. For these functions, there are sets with *positive area* of starting values z_0 for which the sequence $N^n(z_0)$ approaches periodic behavior, and therefore does not converge to any zero of $f(z)$. The structure of these kinds of areas is related to the famous *Mandelbrot set* in fascinating ways.

We seem to have come a long way from an approximation technique in elementary calculus. Yet, of course, it isn't really a long way at all. Perhaps the moral is that in mathematics, there are simple questions ("What happens if we try different starting points for Newton's Method?") that, if we pursue them, lead to surprising and sometimes beautiful answers and new questions.

6. Solutions to the Exercises

1. If $x_n - x_* = c \cdot 10^k$ is correct to k decimal places, then $x_{n+1} - x_* \approx c^2 \cdot 10^{2k}$ will be correct to about $2k$ decimal places.

2. In each case, the last figure is the zero, correct to the shown accuracy:

   a) 1
      1.5
      1.4166 ...
      1.4142156 ...
      1.4142135623746 ...
      1.414213562373095 ...

   b) 1
      2
      1.6
      1.442 ...
      1.4150 ...
      1.4142142 ...
      1.414213562373 ...

   c)   0.7
       −1.294 ...
       −1.43322 ...
       −1.41456 ...
       −1.41421370 ...
       −1.414213562373 ...

   d)   0.5
       −0.2
        0.0085 ...
       −0.0000006165 ...
        0.000000000000 ...

3. a) $N(x) = x/2$.
   b) Each iteration cuts the error in half. Since $(\frac{1}{2})^{10} \approx (\frac{1}{10})^3$, we get that $(0.5)^{50} \approx (0.1)^{15}$. It will take about another 50 iterations to get another 15 decimal places of accuracy. This kind of convergence is called *linear convergence*.

4. The tangent line at $(x_n, f(x_n))$ is level, so it never intersects the x-axis to give x_{n+1}.

5. See **Figure 2** for an example.

6. The Newton sequence jumps about chaotically and never converges:

   $+0.5 \to -0.75 \to +0.29167 \to -1.56845 \to -0.46544 \to$
   $+0.84153 \to -0.17339 \to +2.79697 \to +1.21972 \to +0.19993 \to$
   $-2.40088 \to -0.99218 \to +0.00785 \to -63.710 \to -31.847 \to \cdots$

   We would expect some kind of trouble because $f(x) = 0$ has no real solutions.

7. By symmetry, it is enough to check that for $0 < x < 1/\sqrt{5}$, $N(x) > -x$.

Newton's Method and Fractal Patterns

8. Newton's method fails because the denominator of $N(x)$ becomes undefined at the $(i+1)^{st}$ stage.

11. For $x^2 - 2$, $B(-\sqrt{2}) = (-\infty, 0)$ and $B(\sqrt{2}) = (0, \infty)$. For $x^2 - 4x + 3$, $B(1) = (-\infty, 2)$ and $B(3) = (2, \infty)$. Argue from the geometry of the parabola. In the next section we will generalize to the complex plane the principle that for quadratic polynomials, any initial point is attracted to the root closest to it.

12. $N(z) = -b/a$. The basin $B(-b/a)$ is the entire complex plane.

13. $0.5i \to 1.25i \to 1.025i \to 1.0003049i \to \cdots \to i$.

14. $N(z) = \dfrac{1}{2}\left(z - \dfrac{1}{z}\right) = \dfrac{x}{2}\left(1 - \dfrac{1}{x^2 + y^2}\right) + i\dfrac{y}{2}\left(1 + \dfrac{1}{x^2 + y^2}\right)$.

15. If $y > 0$, then $\dfrac{y}{2}\left(1 + \dfrac{1}{x^2+y^2}\right) > 0$.

16. $z = 1.32472$, $z = -0.662359 \pm 0.56228i$. We found the first root in an earlier example.

17. For example, $(1/2) \pm (1/2)i$ lead to $-(1/2) \pm (\sqrt{3}/2)i$.

References

Becker, Karl-Heinz, and Michael Dörfler. 1989. *Dynamical Systems and Fractals: Computer Experiments in Pascal.* Cambridge: Cambridge University Press.

Cayley, Arthur. 1879. The Newton-Fourier imaginary problem. *American Journal of Mathematics* 2: 97.

Curry, James, Lucy Garnett, and Dennis Sullivan. 1983. On the iteration of a rational function: computer experiments with Newton's method. *Communications in Mathematical Physics* 91: 267–277.

Gleick, James. 1987. *Chaos: Making a New Science.* New York: Viking.

Peitgen, Heinz-Otto, Dietmar Saupe, and Fritz von Haeseler. 1984. Cayley's problem and Julia sets. *Mathematical Intelligencer* 6: 11–20.

Peitgen, Heinz-Otto, and Peter Richter. 1986. *The Beauty of Fractals.* New York: Springer-Verlag.

Saari, Donald, and John Urenko. 1984. Newton's method, circle maps, and chaotic motion. *American Mathematical Monthly* 91: 3–17.

Strang, Gilbert. 1991. A chaotic search for i. *College Mathematics Journal* 22: 3–12.

Acknowledgments

I am grateful to Brent Halsey for generating **Figures 6** and **7**.

This Module was developed in connection with the NSF Calculus Reform in the Liberal Arts College Project, supported by NSF grant USE 8813914. It will be included in the volume *Applications of Calculus* to be issued by that Project.

About the Author

Philip D. Straffin, Jr., received a B.A. from Harvard University, an M.A. from Cambridge University, and a Ph.D. from the University of California—Berkeley in algebraic topology. He has been at Beloit College since 1970 and was chair of Mathematics and Computer Science from 1980 to 1990. He has written two books and a couple of dozen articles on topology, game theory, mathematical political science, and mathematical modeling. He is currently editing a volume, *Applications of Calculus*, as part of an NSF Calculus Reform grant to the Associated College of the Midwest and the Great Lakes College Association.

UMAP

Modules in Undergraduate Mathematics and its Applications

Published in cooperation with the Society for Industrial and Applied Mathematics, the Mathematical Association of America, the National Council of Teachers of Mathematics, the American Mathematical Association of Two-Year Colleges, The Institute of Management Sciences, and the American Statistical Association.

Module 717

3-D Graphics in Calculus and Linear Algebra

Yves Nievergelt

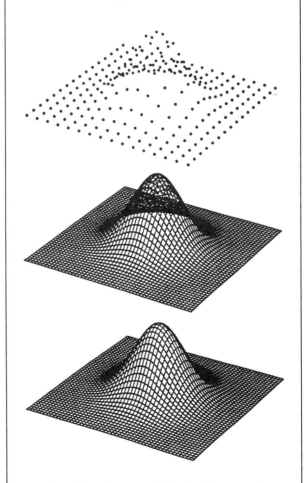

Applications of Calculus and Linear Algebra to 3-D Graphics

COMAP, Inc., Suite 210, 57 Bedford Street, Lexington, MA 02173 (617) 862-7878

INTERMODULAR DESCRIPTION SHEET:	UMAP Unit 717
TITLE:	3-D Graphics in Calculus and Linear Algebra
AUTHOR:	Yves Nievergelt Dept. of Mathematics Eastern Washington University, MS #32 Cheney, WA 99004–2415.
MATHEMATICAL FIELD:	Introductory multivariable calculus and linear algebra
APPLICATION FIELD:	3-D graphics, with or without computers.
TARGET AUDIENCE:	Students in multivariable calculus or in linear algebra
ABSTRACT:	This Module helps students to strengthen their multi-dimensional visual intuition and their ability to solve problems that involve multidimensional concepts in calculus and in linear algebra. The vehicle is an application familiar to students: three-dimensional graphics, studied only with elementary concepts from multivariable calculus and linear algebra. The challenges against which students may measure their progress are to design and modify algorithms for three-dimensional graphics.
PREREQUISITES:	We presuppose a familiarity with the abstract concept of "function" and the particular cases of curves and surfaces in space, with dot and cross products, and with orthonormal bases in the plane and space. Some exercises involve matrix algebra. A familiarity with computers or with supercalculators is necessary only for the reader who wants to verify that the theory applies to the design of 3-D computer graphics on such machines.

©Copyright 1992 by COMAP, Inc. All rights reserved.

COMAP, Inc., Suite 210, 57 Bedford Street, Lexington, MA 02173
(800) 77–COMAP = (800) 772–6627, (617) 862–7878

3-D Graphics in Calculus and Linear Algebra

Yves Nievergelt
Dept. of Mathematics
Eastern Washington University, MS #32
Cheney, WA 99004–2415.

Table of Contents

1. INTRODUCTION 1
2. ORTHOGONAL PROJECTIONS IN SPACE 2
   2.1 The Pythagorean Theorem 2
   2.2 Dot Products and Perpendicularity 3
   2.3 Orthonormal Bases and Orthogonal Projections 6
3. AXONOMETRIES 12
   3.1 The Geometric Concept of "Axonometry" 12
   3.2 Algorithms to Calculate Axonometries 13
4. GRAPHICS 20
   4.1 An Algorithm to Plot Spatial Curves 20
   4.2 An Algorithm to Plot Wire Frames on Graphs 22
5. RENDERING 26
   5.1 "Expanding Horizons" to Remove Hidden Lines 27
   5.2 "Ray Tracing" to Remove Hidden Lines and Shade Surfaces . . 28
6. TERM PROJECTS 30
   6.1 From Computer Programs to Number Theory 30
   6.2 Perspectives without Projective Geometry 30
   6.3 Geometric Perspectives on History 31
   6.4 2-D Graphics in Flatland 31
7. SOLUTIONS TO THE EXERCISES 32
   REFERENCES 39
   ACKNOWLEDGMENTS 40
   ABOUT THE AUTHOR 41

MODULES AND MONOGRAPHS IN UNDERGRADUATE
MATHEMATICS AND ITS APPLICATIONS (UMAP) PROJECT

The goal of UMAP is to develop, through a community of users and developers, a system of instructional modules in undergraduate mathematics and its applications, to supplement existing courses and from which complete courses may be built.

The Project was initially funded by a grant from the National Science Foundation and has been guided by a National Advisory Board of mathematicians, scientists, and educators. UMAP is now supported by the Consortium for Mathematics and Its Applications (COMAP), Inc., a non-profit corporation engaged in research and development in mathematics education.

COMAP Staff

Paul J. Campbell	Editor
Solomon Garfunkel	Executive Director, COMAP
Laurie W. Aragón	Development Director
Philip A. McGaw	Production Manager
Roland Cheyney	Project Manager
Laurie M. Holbrook	Copy Edtior
Dale Horn	Design Assistant
Rob Altomonte	Distribution Coordinator
Sharon McNulty	Executive Assistant

1. Introduction

At the Third Computers & Mathematics Conference, organized by IBM's T.J. Watson Research Center and held at MIT in 1989, speakers emphasized the importance of *multidimensional visualization* in mathematics, the natural sciences, and engineering. Why? Because it is more effective than numerical data in helping to recognize, understand, and communicate patterns. For example, visualizations of such widely different phenomena as the distribution of stress on aircraft fuselages, the evolution of galaxies, and the mathematical homotopy known informally as the "Etruscan Venus," all use five dimensions: the usual spatial three, time for the fourth, and color for coding the fifth.

Meanwhile, even well-prepared undergraduate students encounter inordinate difficulty in visualizing two-dimensional planes in our three-dimensional space. Fortunately, the multidimensional visual intuition that you need to develop in college science and mathematics courses coincides with that used in creating the computer graphics that you see in textbooks and on video screens: lines, planes, projections, dot and cross products, and so forth. Three-dimensional graphics are concrete objects that both *require* mathematics and also *illustrate* mathematics, and to which you may relate in order to internalize the abstract but necessary concepts of mathematics.

Note that "three-dimensional graphics" inherit their name from three-dimensional space, which they represent on a two-dimensional medium, whether a silk screen or a computer screen. Of course, the screen and the graphics themselves have only two dimensions. Thus, the phrase "three-dimensional graphics" refers not only to the resulting pictures, but also to the *methods* to produce two-dimensional pictures of objects from a three-dimensional space.

Our main purpose is to help you to strengthen your multidimensional visual intuition in calculus and in linear algebra, so that you may better understand and solve multivariable problems.

We use 3-D graphics in part because the subject requires no additional background from fields outside of mathematics. Specifically, we focus not on the design or use of three-dimensional graphics, but on their *essence*, that is, on the geometric relationships between three-dimensional spatial objects and their images on a planar screen, described only through elementary concepts from calculus and linear algebra.

The *use* of computer graphics enters the present discussion only as an occasional and optional reality check: to verify that the theory presented here works in practice. To realize the wide applicability of the theory, observe that in this Module, all of the "exhibits" (physical outputs from Hewlett-Packard supercalculators) were produced with machines as small as the Hewlett-Packard HP-28C, while some of the "figures" (which illustrate concepts discussed in the text) were produced with Mathematica on a NeXT machine, and other figures were drawn by hand with only straightedge and compasses.

More advanced topics, such as implementations of commercial algorithms and transformations of surfaces more general than functions, would require both larger machines and a short course on projective geometry. (For a concise treatise of projective geometry and its applications to perspectives, see Hanes [1991].)

2. Orthogonal Projections in Space

An extremely important method for producing three-dimensional graphics employs the orthogonal projections that should be familiar to you from multivariable calculus and linear algebra.

A standard formula expresses orthogonal projections from space onto the plane of a screen in terms of dot products with an orthonormal basis: if the plane of the screen includes a canonical orthonormal basis $(\vec{u}, \vec{v})$, usually with $\vec{u}$ pointing to the right and $\vec{v}$ pointing upward, and if $\vec{x}$ denotes a point in space, then the dot products $\langle \vec{x}, \vec{u} \rangle$ and $\langle \vec{x}, \vec{v} \rangle$ represent the coordinates of the image of the point $\vec{x}$ on the screen.

This formula, though short, constitutes the essence of three-dimensional graphics. We explain and establish it, starting from the Pythagorean theorem, which itself relies upon the properties of distance and area that we observe and take for granted in practice and which we postulate as axioms in geometry.

2.1 The Pythagorean Theorem

For the purpose of three-dimensional graphics, the most important relationship between geometry and algebra consists of the classical theorem of Pythagoras. Denote the usual Euclidean distance by d, so that $d(P, Q)$ stands for the distance from a point P to a point Q.

Theorem 1. (**Pythagorean Theorem**) *Three points A, B, and C form a triangle with a right angle at C if, but only if,*

$$d(A,B)^2 = d(A,C)^2 + d(C,B)^2.$$

In conjunction with a system of cartesian coordinates, the Pythagorean theorem yields the standard formula for the Euclidean distance between two points $P = (x, y)$ and $Q = (u, v)$:

$$d(P,Q) = \sqrt{(x-u)^2 + (y-v)^2}.$$

The concept of distance extends to our three-dimensional space, as does the Pythagorean theorem. The proof in three dimensions follows from two consecutive applications of the theorem in two mutually perpendicular planes,

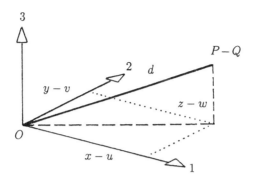

Figure 1. The Pythagorean theorem in three dimensions.

as demonstrated in **Figure 1**. We get the following formula for the Euclidean distance: if $P = (x, y, z)$ and $Q = (u, v, w)$ then

$$d(P,Q) = \sqrt{(x-u)^2 + (y-v)^2 + (z-w)^2}.$$

2.2 Dot Products and Perpendicularity

We now establish the properties of orthogonal projections that are essential to three-dimensional graphics, by means of the concepts of distance and orthogonality, *without the concept of the measure of angles*.

We can exclude measures of angles because the 3-D graphics described here require no angles other than perpendicular lines and because mathematically sound definitions of measures of angles would require analysis (e.g., through integral expressions of the arc length along a circle, or through infinite series for trigonometric functions). For our purposes, distance and perpendicularity will suffice. Also, in what follows, keep in mind that there is no difference between the concepts of "points" and "vectors" in $\mathbb{R}^3$. Traditionally, geometry uses "points" and linear algebra uses "vectors" for essentially the same concepts.

The geometric concept of Euclidean distance in space is related to an algebraic construct known as the dot product.

Definition 1. Let $\vec{p} = (x, y, z)$ and $\vec{q} = (u, v, w)$ represent two points (vectors) in the three-dimensional space $\mathbb{R}^3$. Their *dot product* is the number denoted by $\langle \vec{p}, \vec{q} \rangle$ or by $\vec{p} \cdot \vec{q}$, defined by

$$\langle \vec{p}, \vec{q} \rangle = \langle (x, y, z), (u, v, w) \rangle = xu + yv + zw.$$

Thus, the dot product is a function

$$\langle \, , \, \rangle : \mathbb{R}^3 \times \mathbb{R}^3 \to \mathbb{R}, \quad (\vec{\imath}, \vec{\imath}) \mapsto \langle \vec{\imath}, \vec{\imath} \rangle.$$

EXAMPLE 1. If $\vec{p} = (1, 2, 3)$ and $\vec{q} = (4, 5, 6)$, then

$$\langle \vec{p}, \vec{q} \rangle = \langle (1, 2, 3), (4, 5, 6) \rangle = 1 \cdot 4 + 2 \cdot 5 + 3 \cdot 6 = 32.$$

Exercise

1. Calculate the dot product of $\vec{m} = (2, 3, 6)$ and $\vec{n} = (8, 1, 4)$.

The relationship between the dot product and the Euclidean distance consists of the following fact.

Proposition 1. $d(\vec{p}, \vec{q})^2 = \langle \vec{p} - \vec{q}, \vec{p} - \vec{q} \rangle$.

Proof. With $\vec{p} = (x, y, z)$ and $\vec{q} = (u, v, w)$, the definition of the dot product gives

$$\begin{aligned}
\langle \vec{p} - \vec{q}, \vec{p} - \vec{q} \rangle &= \langle (x, y, z) - (u, v, w), (x, y, z) - (u, v, w) \rangle \\
&= \langle (x - u, y - v, z - w), (x - u, y - v, z - w) \rangle \\
&= (x - u)(x - u) + (y - v)(y - v) + (z - w)(z - w) \\
&= (x - u)^2 + (y - v)^2 + (z - w)^2 \\
&= d(\vec{p}, \vec{q})^2.
\end{aligned}$$

In particular, if $\vec{q}$ lies at the origin ($\vec{q} = \vec{0} = (0, 0, 0)$), then the formula becomes $d(\vec{p}, \vec{0})^2 = \langle \vec{p}, \vec{p} \rangle$, and the length of the segment from $\vec{0}$ to $\vec{p}$ equals $\sqrt{\langle \vec{p}, \vec{p} \rangle}$, a quantity known as the "norm" of $\vec{p}$.

Definition 2. The *norm* of a vector $\vec{p} \in \mathbb{R}^3$ is the length, denoted by $\|\vec{p}\|$, of the segment from the origin to the point $\vec{p}$; thus,

$$\|\vec{p}\| = \sqrt{\langle \vec{p}, \vec{p} \rangle}.$$

EXAMPLE 2. If $\vec{p} = (1, 2, 2)$, then $\langle \vec{p}, \vec{p} \rangle = \langle (1, 2, 2), (1, 2, 2) \rangle = 1 \cdot 1 + 2 \cdot 2 + 2 \cdot 2 = 9$ and

$$\|(1, 2, 2)\| = \sqrt{\langle (1, 2, 2), (1, 2, 2) \rangle} = \sqrt{9} = 3.$$

Exercise

2. Calculate the norm of $\vec{u} = (2, 3, 6)$.

REMARK 1. The dot product is also *symmetric*, which means that

$$\langle \vec{p}, \vec{q} \rangle = \langle \vec{q}, \vec{p} \rangle,$$

and *bilinear*, which means that

$$\langle p\vec{p} + q\vec{q}, r\vec{r} + s\vec{s} \rangle = pr\langle \vec{p}, \vec{r} \rangle + ps\langle \vec{p}, \vec{s} \rangle + qr\langle \vec{q}, \vec{r} \rangle + qs\langle \vec{q}, \vec{s} \rangle$$

for all real numbers $p, q, r, s \in \mathbb{R}$ and for all points (vectors) $\vec{p}, \vec{q}, \vec{r}, \vec{s} \in \mathbb{R}^3$. The proof of the symmetry and bilinearity of the dot product proceeds by expressing each vector with coordinates, then by expressing the dot products by the definition of the dot product, and finally by observing that both sides of the equations are equal.

The following proposition expresses the relation between the dot product and perpendicularity.

Proposition 2. *Three points $\vec{0}$ (the origin), $\vec{p}$, and $\vec{q}$ form a right triangle in space, with a right angle at $\vec{0}$, if, but only if, $\langle \vec{p}, \vec{q} \rangle = 0$.*

Proof. By the Pythagorean theorem,

$$d(\vec{p}, \vec{q})^2 = d(\vec{p}, \vec{0})^2 + d(\vec{0}, \vec{q})^2.$$

Hence, by the preceding proposition, expressing distances with dot products yields

$$\langle \vec{p} - \vec{q}, \vec{p} - \vec{q} \rangle = \langle \vec{p}, \vec{p} \rangle + \langle \vec{q}, \vec{q} \rangle.$$

Multiplying out the left-hand side according to symmetry and bilinearity gives

$$\langle \vec{p} - \vec{q}, \vec{p} - \vec{q} \rangle = \langle \vec{p}, \vec{p} \rangle - 2\langle \vec{p}, \vec{q} \rangle + \langle \vec{q}, \vec{q} \rangle.$$

The two equalities just established lead together to

$$\langle \vec{p}, \vec{p} \rangle - 2\langle \vec{p}, \vec{q} \rangle + \langle \vec{q}, \vec{q} \rangle = \langle \vec{p}, \vec{p} \rangle + \langle \vec{q}, \vec{q} \rangle,$$

and then canceling similar terms and dividing both sides by 2 yields

$$\langle \vec{p}, \vec{q} \rangle = 0.$$

To prove the converse, assume that $\langle \vec{p}, \vec{q} \rangle = 0$; then

$$\begin{aligned}
d(\vec{p}, \vec{q})^2 &= \langle \vec{p} - \vec{q}, \vec{p} - \vec{q} \rangle \\
&= \langle \vec{p}, \vec{p} \rangle - 2\langle \vec{p}, \vec{q} \rangle + \langle \vec{q}, \vec{q} \rangle \\
&= \langle \vec{p}, \vec{p} \rangle - 2 \times 0 + \langle \vec{q}, \vec{q} \rangle \\
&= \langle \vec{p}, \vec{p} \rangle + \langle \vec{q}, \vec{q} \rangle \\
&= d(\vec{p}, \vec{0})^2 + d(\vec{0}, \vec{q})^2.
\end{aligned}$$

Definition 3. Two vectors $\vec{p}$ and $\vec{q}$ in $\mathbb{R}^3$ are *orthogonal*, or *perpendicular*, if, but only if, $\langle \vec{p}, \vec{q} \rangle = 0$, which means that $\vec{p}, \vec{q}$, and $\vec{0}$ form a triangle with a right angle at the origin, $\vec{0}$.

EXAMPLE 3. The vectors $\vec{p} = (1,2,2)$ and $\vec{q} = (2,1,-2)$ are orthogonal because

$$\langle \vec{p}, \vec{q} \rangle = \langle (1,2,2), (2,1,-2) \rangle = 1 \cdot 2 + 2 \cdot 1 + 2 \cdot (-2) = 0.$$

Moreover, observe that

$$d(\vec{p}, \vec{0})^2 = \langle \vec{p}, \vec{p} \rangle = \langle (1,2,2), (1,2,2) \rangle = 1 \cdot 1 + 2 \cdot 2 + 2 \cdot 2 = 9,$$

$$d(\vec{0}, \vec{q})^2 = \langle \vec{q}, \vec{q} \rangle = \langle (2,1,-2), (2,1,-2) \rangle = 2 \cdot 2 + 1 \cdot 1 + (-2) \cdot (-2) = 9,$$

$$\begin{aligned} d(\vec{p}, \vec{q})^2 &= \langle \vec{p} - \vec{q}, \vec{p} - \vec{q} \rangle = \langle (1,2,2) - (2,1,-2), (1,2,2) - (2,1,-2) \rangle \\ &= \langle (-1,1,4), (-1,1,4) \rangle = (-1) \cdot (-1) + 1 \cdot 1 + 4 \cdot 4 = 18; \end{aligned}$$

so that, as asserted by the preceding proposition,

$$d(\vec{p}, \vec{q})^2 = 18 = 9 + 9 = d(\vec{p}, \vec{0})^2 + d(\vec{0}, \vec{q})^2.$$

REMARK 2. Two vectors $\vec{p}$ and $\vec{q}$ are orthogonal if, but only if, $\vec{q}$ and $\vec{p}$ are orthogonal, because $\langle \vec{p}, \vec{q} \rangle = \langle \vec{q}, \vec{p} \rangle$ by symmetry of the dot product (see **Remark 1**). Thus, the relation of orthogonality is symmetric. However, it is not transitive, because there exist vectors $\vec{p}, \vec{q}$, and $\vec{g}$ such that $\vec{p}$ and $\vec{q}$ are orthogonal, $\vec{q}$ and $\vec{g}$ are orthogonal, but $\vec{p}$ and $\vec{g}$ need not be orthogonal. For example, let $\vec{p} = (1,2,2)$, $\vec{q} = (2,1,-2)$, as in **Example 3**, and let $\vec{g} = \vec{p} = (1,2,2)$. Then $\langle \vec{p}, \vec{q} \rangle = 0$, $\langle \vec{q}, \vec{g} \rangle = \langle \vec{q}, \vec{p} \rangle = 0$; but by **Example 2**, $\langle \vec{p}, \vec{g} \rangle = \langle \vec{p}, \vec{p} \rangle = 9 \neq 0$.

Exercise

3. Verify that each pair of distinct vectors among the following three forms a pair of orthogonal vectors:

$$\vec{u} = (2,3,6), \quad \vec{v} = (3,-6,2), \quad \text{and} \quad \vec{w} = (6,2,-3).$$

By **Remark 2**, it suffices to verify that $\vec{u}$ and $\vec{v}$ are orthogonal, that $\vec{v}$ and $\vec{w}$ are orthogonal, and that $\vec{w}$ and $\vec{u}$ are orthogonal.

2.3 Orthonormal Bases and Orthogonal Projections

Three-dimensional graphics involve three perpendicular directions: the direction of the viewer (from the screen to the observer), and the horizontal and vertical directions on the screen. These three perpendicular directions correspond algebraically to an orthonormal basis for the ambient space.

Recall from linear algebra that a basis of $\mathbb{R}^3$ consists of three linearly independent vectors, which span the whole space. For example, the standard (or "canonical") basis consists of

$$\vec{e}_1 = (1,0,0), \quad \vec{e}_2 = (0,1,0), \quad \text{and} \quad \vec{e}_3 = (0,0,1).$$

For each point (vector) $\vec{x} = (x, y, z) \in \mathbb{R}^3$, observe that the coordinates of $\vec{x}$ with respect to the canonical basis equal the dot products of $\vec{x}$ with the elements of the basis:

$$x = x \cdot 1 = \langle (x, y, z), (1, 0, 0) \rangle,$$
$$y = y \cdot 1 = \langle (x, y, z), (0, 1, 0) \rangle,$$
$$z = z \cdot 1 = \langle (x, y, z), (0, 0, 1) \rangle.$$

A similar relationship between coordinates and dot products holds for a special type of basis:

Definition 4. An *orthonormal basis* of $\mathbb{R}^3$ is a basis $(\vec{u}, \vec{v}, \vec{w})$ such that each element of the basis has length 1 and such that all pairs of distinct elements of the basis are mutually perpendicular:

$$\langle \vec{u}, \vec{u} \rangle = \langle \vec{v}, \vec{v} \rangle = \langle \vec{w}, \vec{w} \rangle = 1,$$
$$\langle \vec{u}, \vec{v} \rangle = \langle \vec{v}, \vec{w} \rangle = \langle \vec{w}, \vec{u} \rangle = 0.$$

EXAMPLE 4. The basis consisting of the following three vectors, shown in **Figure 2**, is orthonormal:

$$\vec{p} = (1/3, 2/3, 2/3), \quad \vec{q} = (2/3, 1/3, -2/3), \quad \vec{j} = (2/3, -2/3, 1/3).$$

The verification follows from calculations like those in **Examples 2 and 3**.

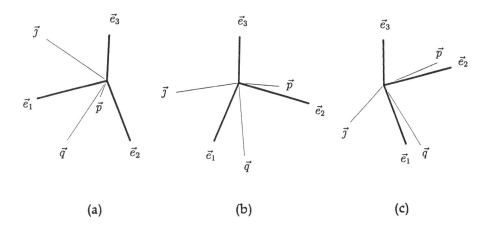

Figure 2 The canonical basis, $(\vec{e}_1, \vec{e}_2, \vec{e}_3)$, and the orthonormal basis $(\vec{p}, \vec{q}, \vec{j})$, viewed from the points: **a.** $(2, 6, 9)$; **b.** $(6, 2, 9)$; and **c.** $(6, -2, 9)$.

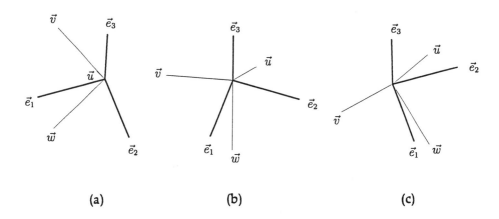

Figure 3 The canonical basis, $(\vec{e}_1, \vec{e}_2, \vec{e}_3)$, and the orthonormal basis $(\vec{u}, \vec{v}, \vec{w})$, viewed from the points: **a.** $(2, 6, 9)$; **b.** $(6, 2, 9)$; and **c.** $(6, -2, 9)$.

Exercise

4. Verify that the following vectors, shown in **Figure 3**, form an orthonormal basis:

$$\vec{u} = (2/7, 3/7, 6/7), \quad \vec{v} = (3/7, -6/7, 2/7), \quad \text{and} \quad \vec{w} = (6/7, 2/7, -3/7).$$

Orthonormal bases simplify the calculation of norms and coordinates, as does the canonical basis.

Proposition 3. *If $(\vec{u}, \vec{v}, \vec{w})$ is an orthonormal basis, and if a, b, and c denote the coordinates of a point $\vec{x} \in \mathbb{R}^3$ with respect to that basis, so that $\vec{x} = a\vec{u} + b\vec{v} + c\vec{w}$, then*

$$a = \langle \vec{x}, \vec{u} \rangle, \quad b = \langle \vec{x}, \vec{v} \rangle, \quad \text{and} \quad c = \langle \vec{x}, \vec{w} \rangle.$$

Proof.

$$\begin{aligned}\langle \vec{x}, \vec{u} \rangle &= \langle a\vec{u} + b\vec{v} + c\vec{w}, \vec{u} \rangle \\ &= a\langle \vec{u}, \vec{u} \rangle + b\langle \vec{v}, \vec{u} \rangle + c\langle \vec{w}, \vec{u} \rangle \\ &= a \cdot 1 + b \cdot 0 + c \cdot 0 = a,\end{aligned}$$

and similarly for the other two coordinates.

EXAMPLE 5. Consider the basis $(\vec{p}, \vec{q}, \vec{\jmath})$ in **Example 4**, and let $\vec{x} = (1, 2, 3)$. Then $\vec{x} = a\vec{p} + b\vec{q} + c\vec{\jmath}$ with

$$a = \langle \vec{x}, \vec{p} \rangle = \langle (1, 2, 3), (1/3, 2/3, 2/3) \rangle = 1 \cdot 1/3 + 2 \cdot 2/3 + 3 \cdot 2/3 = 11/3.$$

Similarly, $b = \langle \vec{x}, \vec{q} \rangle = -2/3$ and $c = \langle \vec{x}, \vec{\jmath} \rangle = 1/3$. As a verification,

$$a\vec{p} + b\vec{q} + c\vec{\jmath} = 11/3(1/3, 2/3, 2/3) - 2/3(2/3, 1/3, -2/3) + 1/3(2/3, -2/3, 1/3) = (1, 2, 3).$$

Exercise

5. Calculate the coordinates of $\vec{x} = (1, 2, 3)$ with respect to the orthonormal basis $(\vec{u}, \vec{v}, \vec{w})$ in **Exercise 4**. Then verify your results.

The following proposition shows that the calculation of the norm, introduced in **Definition 2**, remains the same with every orthonormal basis as it is with the canonical basis.

Proposition 4. *If $(\vec{u}, \vec{v}, \vec{w})$ is an orthonormal basis, and if $\vec{x} = a\vec{u} + b\vec{v} + c\vec{w}$, then $\|\vec{x}\| = \sqrt{a^2 + b^2 + c^2}$.*

Proof. Recall that $\|\vec{x}\| = \sqrt{\langle \vec{x}, \vec{x} \rangle}$ by definition of the norm. Consequently,

$$\|\vec{x}\| = \sqrt{\langle \vec{x}, \vec{x} \rangle} = \sqrt{\langle a\vec{u} + b\vec{v} + c\vec{w}, a\vec{u} + b\vec{v} + c\vec{w} \rangle}.$$

To obtain the desired result, multiply out the dot product, using the bilinearity of the dot product and the orthonormality of the basis.

EXAMPLE 6. With $\vec{x} = (1, 2, 3)$, recall from **Example 5** that $(1, 2, 3) = {}^{11}/_3 \vec{p} - {}^2/_3 \vec{q} + {}^1/_3 \vec{s}$, and that $(\vec{p}, \vec{q}, \vec{s})$, as in **Example 4** is an orthonormal basis. Therefore,

$$\|(1, 2, 3)\| = \sqrt{({}^{11}/_3)^2 + (-{}^2/_3)^2 + ({}^1/_3)^2} = \sqrt{{}^{121}/_9 + {}^4/_9 + {}^1/_9} = \sqrt{{}^{126}/_9} = \sqrt{14}.$$

To verify this result with the canonical basis, observe that

$$\|(1, 2, 3)\| = \sqrt{1^2 + 2^2 + 3^2} = \sqrt{14}.$$

The concepts developed so far now provide the foundations of orthogonal projections.

Definition 5. Let $W \subset \mathbb{R}^3$ represent a two-dimensional plane through the origin in space. The *orthogonal projection* onto the plane W is the function

$$P : \mathbb{R}^3 \to W,$$

defined by the requirement that $P(\vec{x})$ be the point in W that lies closest to $\vec{x}$. In other words, $d(P(\vec{x}), \vec{x}) \leq d(\vec{y}, \vec{x})$ for every $\vec{y} \in W$.

The following proposition provides a means to calculate orthogonal projections by using orthonormal bases. In the context of 3-D graphics, it also provides an algorithm to determine the image on the screen of a point in space, as explained in the next section.

Proposition 5. *Let W represent a two-dimensional linear subspace of $\mathbb{R}^3$ (a plane through the origin), and let $(\vec{u}, \vec{v})$ stand for an orthonormal basis of W. Also, let $P : \mathbb{R}^3 \to W$ denote the orthogonal projection onto W. Then the coordinates of $P(\vec{x})$ with respect to the orthonormal basis $(\vec{u}, \vec{v})$ are the dot products of $\vec{x}$ with the basis:*

$$P(\vec{x}) = \langle \vec{x}, \vec{u} \rangle \cdot \vec{u} + \langle \vec{x}, \vec{v} \rangle \cdot \vec{v}.$$

Proof. By choosing a vector $\vec{w}$ with length 1 and perpendicular to the plane W, complete the basis of W to an orthonormal basis $(\vec{u}, \vec{v}, \vec{w})$ of the whole space $\mathbb{R}^3$. Then

$$\vec{x} = \langle \vec{x}, \vec{u} \rangle \cdot \vec{u} + \langle \vec{x}, \vec{v} \rangle \cdot \vec{v} + \langle \vec{x}, \vec{w} \rangle \cdot \vec{w}.$$

Also, consider any point $\vec{y} \in W$; since $(\vec{u}, \vec{v})$ forms a basis of W, there exist real numbers h and k such that $\vec{y} = h\vec{u} + k\vec{v}$. Hence,

$$d(\vec{x}, \vec{y}) = d(\vec{x} - \vec{y}, \vec{0}) = \|\vec{x} - \vec{y}\|$$

$$= \sqrt{(\langle \vec{x}, \vec{u} \rangle - h)^2 + (\langle \vec{x}, \vec{v} \rangle - k)^2 + (\langle \vec{x}, \vec{w} \rangle)^2} \geq \sqrt{\langle \vec{x}, \vec{w} \rangle^2}.$$

Observe that $d(\vec{x}, \vec{y})$ reaches a minimum value, $|\langle \vec{x}, \vec{w} \rangle|$, if, but only if, $h = \langle \vec{x}, \vec{u} \rangle$ and $k = \langle \vec{x}, \vec{v} \rangle$. This means that $P(\vec{x}) = h\vec{u} + k\vec{v}$ with $h = \langle \vec{x}, \vec{u} \rangle$ and $k = \langle \vec{x}, \vec{v} \rangle$.

EXAMPLE 7. If $\vec{x} = (1, 2, 3)$, and if W represents the plane spanned by the orthonormal basis $(\vec{p}, \vec{q})$ as in **Example 4**, then

$$P((1, 2, 3)) = {}^{11}\!/_3(1/3, 2/3, 2/3) - 2/3(2/3, 1/3, -2/3) = (7/9, 20/9, 26/9)$$

with both dot products, $11/3$ and $-2/3$, as calculated in **Example 5**. Observe that the *calculation* of the projection by using the formula does not require any knowledge of the vector $\vec{w}$ perpendicular to W; only the above *proof* of the formula for the projection requires $\vec{w}$.

Exercise

6. Calculate the orthogonal projection of $\vec{x} = (1, 2, 3)$ onto the plane W spanned by the orthonormal basis $(\vec{u}, \vec{v})$ with $\vec{u} = (2/7, 3/7, 6/7)$ and $\vec{v} = (3/7, -6/7, 2/7)$ (see **Exercises 4** and **5**).

Given an orthonormal basis $(\vec{u}, \vec{v})$ for a plane $W \subset \mathbb{R}^3$, there exist two choices for the vector $\vec{w}$ with length 1 and perpendicular to W, expressed by the *cross product* $\vec{w} = \pm \vec{u} \times \vec{v}$. By definition,

$$(x, y, z) \times (u, v, w) = (yw - zv, zu - xw, xv - yu).$$

EXAMPLE 8. Consider the vectors $\vec{p} = (1, 2, 2)$ and $\vec{q} = (2, 1, -2)$; then

$$\vec{p} \times \vec{q} = (1, 2, 2) \times (2, 1, -2) = (2 \cdot (-2) - 2 \cdot 1, 2 \cdot 2 - 1 \cdot (-2), 1 \cdot 1 - 2 \cdot 2)$$

$$= (-6, 6, -3) = -3(2, -2, 1).$$

REMARK 3. (VECTOR ALGEBRA WITH SUPERCALCULATORS) Supercalculators offer built-in commands to compute dot products, cross products, and norms, as demonstrated in **Table 1**.

3-D Graphics in Calculus and Linear Algebra

Table 1.

Vector algebra with the HP-28C&S and HP-48SX.

The commands DOT, CROSS, and ABS are on the third page of the ARRAY menu on the HP-28C&S, and in the VECTR submenu of the MTH menu on the HP-48SX.

Keys	Comments	Display
[1 2 2] ABS	Compute the Euclidean norm.	3
[1 2 2] ENTER [2 1 -2] DOT	Compute the dot product.	0
[1 2 2] ENTER [2 1 -2] CROSS	Compute the cross product.	[-6 6 -3]

Routine Exercises

Perform the following computations by hand or with a supercalculator.

7. Calculate the cross product $(4, 1, 8) \times (4, -8, -1) = (?, ?, ?)$.

8. Verify that the following three vectors form an orthonormal basis of $\mathbb{R}^3$:

$$\vec{h} = (4/9, 1/9, 8/9), \quad \vec{k} = (4/9, -8/9, -1/9), \quad \text{and} \quad \vec{\ell} = (7/9, 4/9, -4/9).$$

9. Compute the coordinates of the vector $\vec{x} = (1, 2, 3)$ with respect to the orthonormal basis $(\vec{h}, \vec{k}, \vec{\ell})$ in the preceding exercise. Verify your results. (Does $\vec{x} = a\vec{h} + b\vec{k} + c\vec{\ell}$?)

10. Compute the orthogonal projection of the vector $\vec{x} = (1, 2, 3)$ onto the two-dimensional plane W spanned by the vectors $\vec{h}$ and $\vec{k}$ (as in the preceding exercises).

11. Verify that the two vectors

$$\vec{r} = (2/11, 6/11, 9/11), \quad \text{and} \quad \vec{s} = (9/11, -6/11, 2/11)$$

are orthonormal (orthogonal and with length equal to 1); then calculate a third vector, $\vec{t} = (?, ?, ?)$, such that $\vec{t}$ has length 1 and is perpendicular to both $\vec{r}$ and $\vec{s}$.

Theoretical Exercises

12. Prove that for every pair of vectors $\vec{x}$ and $\vec{y}$ in $\mathbb{R}^3$, the cross product $\vec{x} \times \vec{y}$ is perpendicular to both $\vec{x}$ and $\vec{y}$.

13. (For readers familiar with Lagrange multipliers) Consider a plane W through the origin, perpendicular to a nonzero vector $\vec{w} = (m, r, s)$. Also, let $\vec{x_0} = (x_0, y_0, z_0) \in \mathbb{R}^3$ represent any point in space. For each point $\vec{x} = (x, y, z) \in W$, denote by $f(x, y, z) = d(\vec{x_0}, \vec{x})^2$ the square of the distance from $\vec{x_0}$ to the point $\vec{x} \in W$. Write the system of equations that results from minimizing f, subject to the constraint that $\vec{x} \in W$. Then provide a verbal interpretation of the system.

3. Axonometries

3.1 The Geometric Concept of "Axonometry"

Three-dimensional graphics represent objects in the ambient 3-D space on a 2-D medium, for instance, a silk screen or a computer screen, in order to display an observer's view of spatial objects. To this end, a three-dimensional graphics procedure involves a function $G : \mathbb{R}^3 \to \mathbb{R}^2$ mapping a model of the ambient space, $\mathbb{R}^3$, onto a model of the screen, $\mathbb{R}^2$. Examples of such functions are celestial maps (stereographic projections), architectural perspectives (central projections), and technical axonometries (orthogonal parallel projections). Among graphics projections, axonometries—despite their forbiddingly unfamiliar name!—may be the simplest kind in the context of calculus and linear algebra, because they involve only a few operations from vector algebra.

To understand the role of vector algebra in an axonometric 3-D graphics procedure, imagine an observer at a site $\vec{s}$ far away from the origin in $\mathbb{R}^3$, as in **Figure 4a**, who is looking toward the origin, in the direction $-\vec{w} = -(1/\|\vec{s}\|)\vec{s}$ (so that $\vec{w}$ denotes a unit vector pointing toward the observer). An approximation of the observer's view of the ambient space consists of the image of the orthogonal projection $P : \mathbb{R}^3 \to W = \vec{w}^\perp$ onto the two-dimensional subspace W perpendicular to the direction of sight $-\vec{w}$, as in **Figure 4b**. Technically, such a projection bears the name of "axonometry."

In terms of calculus, the concept of axonometry means that for each point $\vec{r}$ in the ambient space $\mathbb{R}^3$, the projection $P(\vec{r})$ is the point closest to $\vec{r}$ on the screen W. In terms of linear algebra, the concept of axonometry means that $P : \mathbb{R}^3 \to W$ is a linear transformation that is a projection (so that that P composed with itself is just P again, $P \circ P = P$), and that the kernel of P is orthogonal to the range of P. (Recall that the kernel, or null space, of P consists of all the vectors that P maps to zero.)

Definition 6. An *axonometry* is a linear orthogonal projection $P : \mathbb{R}^3 \to W$ onto a two-dimensional linear subspace of $\mathbb{R}^3$.

(To render all three canonical vectors recognizably on the screen, some authors also require that the kernel of an axonometry not be parallel to any vector in the canonical basis.)

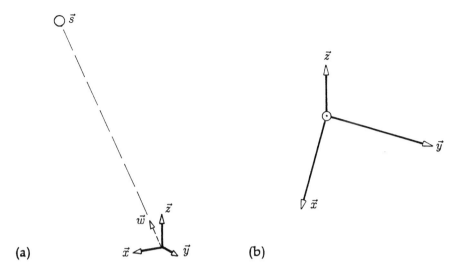

Figure 4 a. What picture of the canonical basis does an observer see from $\vec{s}$? **b.** The observer sees the orthogonal projection of the basis on a plane perpendicular to $\vec{s}$. (For this particular drawing, $\vec{s} = (4, 1, 8)$.)

3.2 Algorithms to Calculate Axonometries

In practice, plotting the axonometric images of points on a computer screen requires a coordinate system, to address individual picture elements called "pixels." To this effect, choose right-handed orthonormal bases, $(\vec{x}, \vec{y}, \vec{z})$ for the ambient space $\mathbb{R}^3$, and $(\vec{u}, \vec{v})$ for the screen $W \approx \mathbb{R}^2$, subject to two conditions:

- to the observer at $\vec{s} = \|\vec{s}\|\vec{w}$, the vector $\vec{u}$ must appear horizontal and pointing toward the right, and

- $\vec{v}$ must appear vertical and pointing upward, as in **Figure 5a**.

Algebraically, these conditions mean that $(\vec{u}, \vec{v}, \vec{w})$ is another right-handed orthonormal basis of $\mathbb{R}^3$ (as in **Figure 5b**), such that

- $\langle P(\vec{z}), \vec{u} \rangle = 0$, so that the vertical component of $\vec{u}$ vanishes; and

- $\langle P(\vec{z}), \vec{v} \rangle > 0$, so that $\vec{v}$ points upward.

In this context, determining the image (u, v) on the screen $W \approx \mathbb{R}^2$ of a point $\vec{r} = (x, y, z)$ in $\mathbb{R}^3$ amounts to calculating the coordinates u and v of $P(\vec{r})$ relative to the basis $(\vec{u}, \vec{v})$. The calculation requires the solution of the following two problems.

Problem 1. Relative to the basis $(\vec{x}, \vec{y}, \vec{z})$, given a vector $\vec{s}$ in the direction of the observer, calculate the coordinates of the vectors just described, viz.,

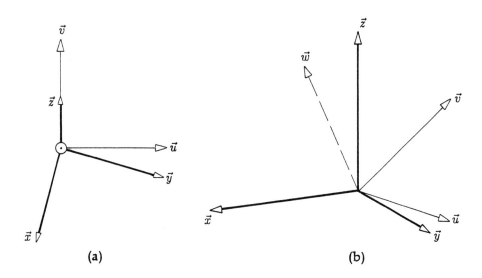

Figure 5 a. The screen lies in the plane of the picture, with an orthonormal basis $(\vec{u}, \vec{v})$ to address individual pixels. **b.** The vectors $\vec{u}$ and $\vec{v}$ on the screen and the vector $\vec{w}$ toward the observer form another right-handed orthonormal basis of $\mathbb{R}^3$.

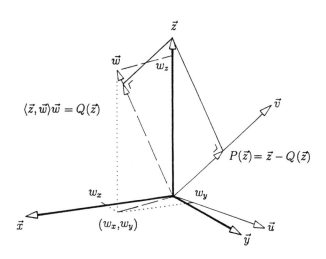

Figure 6 Construction of $\vec{v}$ and then $\vec{u}$. Incidentally, notice the congruent triangles, which imply that the length of the projection $P(\vec{z})$ equals the length of the projection of $\vec{w}$ on the plane spanned by $(\vec{x}, \vec{y})$.

$\vec{u} = (u_x, u_y, u_z)$ and $\vec{v} = (v_x, v_y, v_z)$, by means of vector or matrix algebra—that is, by using the operations $+$, $-$, $\times$, $\langle\ \rangle$, $\|\ \|$, and $\cdot$

Solution 1. (See **Figure 6**.) First, notice that if $\vec{v}$ appears vertical upward to the observer, then $\vec{v}$ lies in the direction of the projection of $\vec{z}$ on the screen. The fact that $\vec{v}$ must be parallel to $P(\vec{z})$ also follows from the requirements that $\vec{u}$ be orthogonal to both $\vec{v}$ and $P(\vec{z})$, whence $\langle P(\vec{z}), \vec{u}\rangle = 0$ and $\langle \vec{v}, \vec{u}\rangle = 0$. Thus, $\vec{v} = (1/\|P(\vec{z})\|)\,P(\vec{z})$. To calculate $P(\vec{z})$, consider the complementary orthogonal projection Q, which maps $\mathbb{R}^3$ onto the kernel of P, here spanned by $\vec{w}$. (Q has the advantage over P that its range already has a given orthonormal basis, $\{\vec{w}\}$, whereas we do not yet know $(\vec{u},\vec{v})$. The equality $P = I - Q$ will then yield a formula for P.) From standard properties of scalar products,

$$Q : \mathbb{R}^3 \to \mathrm{Span}\{\vec{w}\},\ \vec{r} \mapsto \langle \vec{r}, \vec{w}\rangle \vec{w}.$$

Next, from the properties of complementary orthogonal projections, $P + Q = I$. Hence, $P = I - Q$ and the following expression yields formulae for $P(\vec{r})$, hence for $P(\vec{z})$ and $\vec{v}$.

$$P(\vec{r}) = I(\vec{r}) - Q(\vec{r}) = \vec{r} - \langle \vec{r}, \vec{w}\rangle \vec{w}.$$

ALGORITHM 1: Calculation of the basis $(\vec{u}, \vec{v}, \vec{w})$. Given any vector $\vec{s} = (s_x, s_y, s_z)$ (but not a multiple of $\vec{z}$) pointing toward the observer, perform the following operations.

$$\vec{w} = (1/\|\vec{s}\|)\,\vec{s},$$
$$P(\vec{z}) = \vec{z} - \langle \vec{z}, \vec{w}\rangle \vec{w},$$
$$\vec{v} = (1/\|P(\vec{z})\|)\,P(\vec{z}),$$
$$\vec{u} = \vec{v} \times \vec{w}.$$

EXAMPLE 9. Suppose that $\vec{s} = (3, 4, 12)$. Then $\|\vec{s}\|^2 = 3^2 + 4^2 + 12^2 = 169 = 13^2$; thus, $\|\vec{s}\| = 13$ and

$$\vec{w} = (1/\|\vec{s}\|)\,\vec{s} = (1/13)\,(3, 4, 12) = (3/13, 4/13, 12/13).$$

Next,

$$P(\vec{z}) = \vec{z} - \langle \vec{z}, \vec{w}\rangle \vec{w} = (0, 0, 1) - (12/13)(3/13, 4/13, 12/13) = (1/169)(-36, -48, 25).$$

As above for $\vec{s}$, arithmetic shows that $\|P(\vec{z})\| = 65/169$, and, consequently,

$$\vec{v} = (1/\|P(\vec{z})\|)\,P(\vec{z}) = (1/65)(-36, -48, 25) = (-36/65, -48/65, 25/65).$$

Finally,

$$\vec{u} = \vec{v} \times \vec{w} = (v_y w_z - v_z w_y,\ v_z w_x - v_x w_z,\ v_x w_y - v_y w_x) = \ldots = (-4/5, 3/5, 0).$$

Exercise

14. Suppose that $\vec{s} = (9, 12, 20)$. Calculate the orthonormal basis $(\vec{u}, \vec{v}, \vec{w})$, as in **Example 9**; give exact results. (For ease of calculation and verification, all results in this particular exercise have a finite decimal expansion with at most two nonzero digits.)

An alternative method for constructing the basis $(\vec{u}, \vec{v}, \vec{w})$ appears in the exercises.

With the orthonormal basis $(\vec{u}, \vec{v}, \vec{w})$ constructed, including the basis $(\vec{u}, \vec{v})$ for the screen and the direction $\vec{w}$ perpendicular to the screen and toward the observer, there remains the problem of determining the image on the screen of a point in space.

Problem 2. Given an orthonormal basis $(\vec{u}, \vec{v}, \vec{w})$ and the coordinates (x, y, z) of a point $\vec{r}$ with respect to the canonical basis $(\vec{x}, \vec{y}, \vec{z})$, calculate the coordinates (u, v) of the orthogonal projection $P(\vec{r})$ of $\vec{r}$ on the plane spanned by $(\vec{u}, \vec{v})$.

Solution 2. As with any orthonormal basis, and as described in **Proposition 5**,

$$P(\vec{r}) = \langle \vec{r}, \vec{u} \rangle \vec{u} + \langle \vec{r}, \vec{v} \rangle \vec{v} + \vec{0}.$$

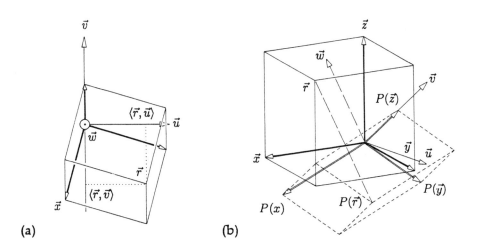

Figure 7 a. To find the coordinates (u, v) of the pixel corresponding to the image of a point $\vec{r} = (x, y, z)$, calculate two scalar products: $u = \langle \vec{r}, \vec{u} \rangle$ and $v = \langle \vec{r}, \vec{v} \rangle$. **b.** To plot an object such as the unit cube, plot points along selected curves on the object, for example, along edges.

ALGORITHM 2: Calculation of the image of a point on the screen. Set $u = \langle \vec{r}, \vec{u} \rangle$ and $v = \langle \vec{r}, \vec{v} \rangle$. The pixel at coordinates (u, v) is the image of $\vec{r}$ (see **Figure 7a**).

3-D Graphics in Calculus and Linear Algebra   145

EXAMPLE 10. Consider the orthonormal basis $(\vec{u}, \vec{v}, \vec{w})$ constructed in **Example 9**, and suppose that $\vec{r} = (1, 2, 3)$, for example. Then

$$\begin{aligned} u &= \langle \vec{r}, \vec{u} \rangle = \langle (1,2,3), (-4/5, 3/5, 0) \rangle \\ &= 1 \times {-4}/5 + 2 \times 3/5 + 0 = 2/5, \\ v &= \langle \vec{r}, \vec{v} \rangle = \langle (1,2,3), (-36/65, -48/65, 25/65) \rangle \\ &= 1 \times {-36}/65 + 2 \times {-48}/65 + 3 \times 25/65 = -57/65. \end{aligned}$$

Thus, the image of $\vec{r} = (1, 2, 3)$ is the pixel at coordinates $(u, v) = (2/5, -57/65)$.

Exercise

15. With the orthonormal basis $(\vec{u}, \vec{v}, \vec{w})$ found in **Exercise 14**, calculate the image of the point $\vec{r} = (1, 2, 3)$, as in **Example 10**.

Applied to a collection of points on an object in space, **Algorithm 2** produces a picture of that object on the screen. However, plotting points alone may merely produce a blob on the screen, as in **Figure 8a**. Instead, plotting points along selected curves on the object, for instance, along edges, as in **Figure 7b**, or on a grid of curves, as in **Figure 8b** or **c**, produces a more representational picture, as explained in more detail in the next sections.

Summary. To plot the axonometry of an object in the ambient three-dimensional space on a two-dimensional screen, choose a vector $\vec{s}$ in the direction of an observer (but not a multiple of $\vec{z}$), and perform the following two steps.

- STEP 1. Only once, before plotting the image of the object, compute the orthonormal basis $(\vec{u}, \vec{v}, \vec{w})$, as in **Algorithm 1** and in **Example 9**.

- STEP 2. For each point $\vec{r}$ to be plotted, compute the coordinates of the image $P(\vec{r})$, as in **Algorithm 2** and **Example 10**.

Exercises

For the following exercises, as in the text, the axonometry P projects $\mathbb{R}^3$ along the direction $\vec{w}$ orthogonally onto the normal plane, spanned by the orthonormal basis $(\vec{u}, \vec{v})$.

Routine Exercises

16. Let $\vec{u} = (u_x, u_y, u_z)$ and $\vec{v} = (v_x, v_y, v_z)$ denote an orthonormal basis on the screen, and let (u, v) represent the coordinates of the projection on the screen, $P(\vec{t})$ of a point $\vec{t} = (x, y, z)$ in space. Express the coordinates u and v in terms of $x, y, z, u_x, u_y, u_z, v_x, v_y,$ and v_z.

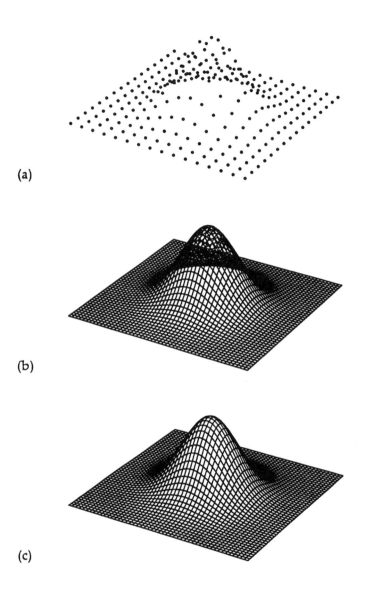

Figure 8 a. Plotting isolated points may fail to create a representational picture of the underlying surface. **b.** Plotting a network of curves may produce a sharper image, as explained in the following section. **c.** Omitting the parts of curves hidden from the observer by an opaque surface may yield a more satisfactory picture, as explained in a subsequent section.

17. Suppose that $\vec{s} = (75, 180, 1456)$. Calculate the orthonormal basis $(\vec{u}, \vec{v}, \vec{w})$. Then calculate the coordinates u and v on the screen of the image of the point $(1, 2, 3)$ in space. State your answers with rational coordinates, in the form p/q for each coordinate. (In this particular exercise, note that although all three vectors $\vec{u}, \vec{v}$, and $\vec{w}$ have only rational coordinates, they involve unwieldy numbers. Therefore, you may want to use your computer or supercalculator to discover and verify simplifications as you perform the calculations.)

18. Among other activities, an applied mathematician's work may include deciphering and understanding someone else's undocumented algorithms, to improve them, or to assert that they perform as intended. For example, explain the role of the following algorithm, in particular, explain how this algorithm relates to three-dimensional graphics:

    data: $H \in \mathbb{R}^3$ (the data consist of one point [vector] in space)
        begin
            let $K := (0, 0, 1) \times H$ (the symbol $\times$ represents cross product)
            let $L := H \times K$
        end
    output: K and L (two additional points [vectors] in space)

Theoretical Exercises

19. Write the matrix of the orthogonal projection P relative to the bases $(\vec{u}, \vec{v}, \vec{w})$ for $\mathbb{R}^3$ and $(\vec{u}, \vec{v})$ for the screen (find its entries).

20. Write the matrix of the same axonometry P relative to the bases $(\vec{x}, \vec{y}, \vec{z})$—the canonical basis of $\mathbb{R}^3$—and $(\vec{u}, \vec{v})$. (Express its rows or columns in terms of the vectors in the bases.) Then apply the result to find the numerical entries of the matrix in the particular case in which, as in **Example 9**,

$$\vec{u} = (-4/5, 3/5, 0), \quad \text{and} \quad \vec{v} = (-36/65, -48/65, 25/65).$$

21. Let $i : W \hookrightarrow \mathbb{R}^3$, $p \mapsto p$ denote the inclusion of the screen plane W into $\mathbb{R}^3$, and consider the composition $i \circ P : \mathbb{R}^3 \to \mathbb{R}^3$, $\vec{r} \mapsto P(\vec{r})$. Express the matrix of $i \circ P$, relative to the canonical basis $(\vec{x}, \vec{y}, \vec{z})$ in both the domain and the range, as a product of two matrices. (The result of this exercise shows the complications arising from the use of canonical bases, as opposed to the simplification afforded by bases related to the projection P.)

22. What problems would arise if $\vec{s}$ were a scalar multiple of $\vec{z}$ (a situation excluded by **Algorithm 1** and the summary)?

Programming Exercises

Complete these programming assignments on the HP-28, HP-48, or any other machine with arithmetic and storage units linked to a graphic display screen.

23. Write a program that, given a nonzero vector $\vec{s}$, computes the orthonormal basis $(\vec{u}, \vec{v}, \vec{w})$ according to **Algorithm 1**. Then test your program on **Example 9** and **Exercises 14** and **17**.

24. Write a program that, given the orthonormal basis $(\vec{u}, \vec{v}, \vec{w})$ and a point $\vec{r} = (x, y, z)$, computes the coordinates u and v of $P(\vec{r})$ according to **Algorithm 2**. Then test your program on **Example 10** and **Exercises 15** and **17**.

25. Combine the results of the preceding two exercises into a program that, given a nonzero vector $\vec{s}$ and a point $\vec{r}$, displays on the screen a picture of the canonical basis and the point $\vec{r}$, all viewed from the direction of $\vec{s}$.

4. Graphics

4.1 An Algorithm to Plot Spatial Curves

The preceding section explained the plotting of isolated points in a three-dimensional space, with suggestions on the use of points as building elements for plotting other spatial objects. One example of spatial objects, which lends itself readily to computer graphics, is a curve in space.

Definition 7. A *curve* in space is a continuous function

$$\gamma : I \to \mathbb{R}^3, \qquad t \mapsto \gamma(t)$$

from a real interval $I \subset \mathbb{R}$ into the three-dimensional space $\mathbb{R}^3$.

EXAMPLE 11. Consider two points in space, $\vec{p} = (p_x, p_y, p_z)$ and $\vec{q} = (q_x, q_y, q_z)$. The function

$$\gamma : [0, 1] \to \mathbb{R}^3, \qquad t \mapsto \gamma(t) = (1-t)\begin{pmatrix} p_x \\ p_y \\ p_z \end{pmatrix} + t\begin{pmatrix} q_x \\ q_y \\ q_z \end{pmatrix}$$

maps the unit interval, $I = [0, 1]$, onto the straight line segment from $\vec{p}$ through $\vec{q}$. For example, if $\vec{p} = (1, 2, 3)$ and $\vec{q} = (2, 4, 8)$, then

$$\gamma : [0, 1] \to \mathbb{R}^3, \qquad t \mapsto (1-t)\begin{pmatrix} 1 \\ 2 \\ 3 \end{pmatrix} + t\begin{pmatrix} 2 \\ 4 \\ 8 \end{pmatrix} = \ldots = \begin{pmatrix} 1+t \\ 2+2t \\ 3+5t \end{pmatrix}.$$

Although a digital computer cannot, of course, handle the infinitely many points on the image of a curve in space, a graphics program that includes a finite but sufficiently large number of points can nevertheless produce the illusion of a continuous curve. To achieve this effect, select a large enough

3-D Graphics in Calculus and Linear Algebra   149

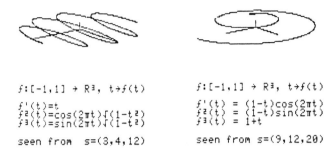

```
f:[-1,1] → R³, t→f(t)              f:[-1,1] → R³, t→f(t)

f¹(t)=t                            f¹(t) = (1-t)cos(2πt)
f²(t)=cos(2πt)√(1-t²)              f²(t) = (1-t)sin(2πt)
f³(t)=sin(2πt)√(1-t²)              f³(t) = 1+t

seen from  s=(3,4,12)              seen from  s=(9,12,20)
```

Exhibit 1. Examples of curves in space, produced by the HP-28C with the program listed below in this subsection. The information printed along with the graphics suggests the kind of data that you may want to record immediately after performing such a graphic experiment, in order to be able to reproduce it later.

integer k, depending upon the desired resolution and the allowed computational time, and choose k values $t_1 < t_2 < \ldots < t_{k-1} < t_k$ in the interval I. Then compute the corresponding k points on the curve γ,

$$\gamma(t_1), \gamma(t_2), \ldots, \gamma(t_{k-1}), \gamma(t_k)$$

and plot each point $\gamma(t_j)$ as explained in the preceding section. In response, the computer produces such graphics as those in **Exhibit 1**. **Exhibit 2** lists programs for the HP-28C/28S and HP-48SX, together with a short tutorial to test them.

EXAMPLE 12. As in the previous example, consider the curve

$$\gamma : [0,1] \to \mathbb{R}^3, \ t \mapsto \gamma(t) = (1-t)\begin{pmatrix}1\\2\\3\end{pmatrix} + t\begin{pmatrix}2\\4\\8\end{pmatrix} = \begin{pmatrix}1+t\\2+2t\\3+5t\end{pmatrix}.$$

To plot the curve γ, select a positive integer, for example, $k = 1001$, and choose 1001 values $t_1 < t_2 < \ldots t_{1000} < t_{1001}$ in $[0,1]$, for instance,

$$t_1 = 0, \ t_2 = 0.001, \ \ldots, \ t_{1000} = 0.999, \ t_{1001} = 1.$$

Next, let the computer calculate the corresponding points,

$$\gamma(0) = \begin{pmatrix}1\\2\\3\end{pmatrix}, \ \gamma(0.001) = \begin{pmatrix}1.001\\2.002\\3.005\end{pmatrix}, \ \ldots, \ \gamma(0.999) = \begin{pmatrix}1.999\\3.998\\7.995\end{pmatrix}, \ \gamma(1) = \begin{pmatrix}2\\4\\8\end{pmatrix}.$$

(Observe that γ, as constructed, starts at $\vec{p}$ and ends at $\vec{q}$.) Finally, or concurrently with the foregoing computations, let the computer plot the images of all 1001 points through a predetermined axonometry.

Exercise

26. Let $\gamma : I \to \mathbb{R}^3$, $t \mapsto (\gamma_x(t), \gamma_y(t), \gamma_z(t))$ represent a curve in space, and let $p = P \circ \gamma$ denote its projection on the screen (through some specified axonometry P). Express the coordinates $p_u(t)$ and $p_v(t)$ of $p(t) = (p_u(t), p_v(t))$, with respect to the basis $(\vec{u}, \vec{v})$, in terms of γ_x, γ_y, γ_z, u_x, u_y, u_z, v_x, v_y, and v_z.

4.2 An Algorithm to Plot Wire Frames on Graphs

Points and curves in space, plotted as described in the preceding subsections, may serve to outline two-dimensional surfaces in a three-dimensional space. In particular, surfaces that coincide with the graph of a function of two variables provide an introduction to computer graphics, without the complications associated with general surfaces. In principle, they also provide enough generality, because, according to the implicit function theorem, each surface consists of patches mapped by functions of two variables. The process of plotting a function of two variables furnishes a concrete illustration of the equivalence between the following three expressions of the concept of "function."

Expression A. A *function* of two variables is a rule, denoted by $f : D \to \mathbb{R}$, which assigns one number, $f(x, y)$, to each point (x, y) in a prespecified subset of the plane, $D \subset \mathbb{R}^2$.

Expression B. A *function* of two variables, denoted by $f : D \to \mathbb{R}$, is a set of ordered pairs $((x, y), z) \in D \times \mathbb{R}$ that consists of one pair $((x, y), z)$ for each point (x, y) in $D \subset \mathbb{R}^2$.

Expression C. A *function* of two variables, denoted by $f : D \to \mathbb{R}$, is a surface (subset) $f \subset \mathbb{R}^3$ that consists of one point (x, y, z) for each point (x, y) in $D \subset \mathbb{R}^2$.

In **Expression A**, the word "rule" means "algorithm," corresponding to the notion of a computer program that computes an output $f(x, y)$ for each input (x, y) from D. If the program computes such values and records them in arrays of the form $((x, y), f(x, y))$ or $(x, y, f(x, y))$, then it produces a collection of points, as described in **Expression B**. Together, these points also form a surface lying above D in space (as in **Expression C**), which a computer graphics procedure may display on a screen. Notice that as a consequence of the equivalence of the three definitions, there exists no distinction between a function and its graph: a function *is* its graph.

Naturally, merely plotting points scattered at random on a surface may result in a picture not of the surface but of chaos. Instead of these scattered

3-D Graphics in Calculus and Linear Algebra

```
------------------------         ------------------------
SPACE CURVES ON HP-28C&S         SPACE CURVES FOR HP-48SX
------------------------         ------------------------
curve                            curve
« basis snug dots rays {         « basis snug dots rays {
curve } ORDER PRLCD »            curve } ORDER GRAPH
                                 »
basis                            basis
« 'W' W ABS INV STO∓ 0 0         « 'W' W ABS INV STO∓ 0 0
1 { 3 } →ARRY DUP W DOT          1 →V3 DUP W DOT W ∓ -
W ∓ - DUP ABS INV ∓ 'V'          DUP ABS INV ∓ 'V' STO V
STO V W CROSS 'U' STO »          W CROSS 'U' STO
                                 »
snug                             snug
« CLΣ b a - 32 / 'h' STO         « CLΣ b a - 128 / 'h'
a b FOR t t X t Y t Z            STO a b
map { 2 } →ARRY Σ+ h               FOR t t X t Y t Z map
STEP SCLΣ 1.0625 DUP ∓H          →V2 Σ+ h
∓W »                               STEP SCATRPLOT 1.0625
                                 DUP ∓H ∓W ERASE
dots                             »
« CLΣ CLLCD b a - 1024 /         dots
'h' STO a b FOR t t X t          « CLΣ b a - 1024 / 'h'
Y t Z ink h STEP »               STO a b
                                   FOR t t X t Y t Z ink
rays                             h
« 64 INV 'h' STO 0 1 FOR           STEP
t t 0 0 ink 0 t 0 ink 0          »
0 t ink h STEP »
                                 rays
ink                              « 64 INV 'h' STO 0 1
« map R→C PIXEL »                  FOR t t 0 0 ink 0 t 0
                                 ink 0 0 t ink h
map                                STEP
« { 3 } →ARRY DUP U DOT          »
SWAP V DOT »
------------------------         ink
EXAMPLE: a=-2π ≤t≤ 2π=b          « map R→C PIXON
 t → (cos(t),sin(t),t)           »
seen from (75,180,1456)
                                 map
(1) Store  a  and  b:            « →V3 DUP U DOT SWAP V
                                 DOT
 π →NUM 2 ∓ ENTER CHS            »
'a' STO 'b' STO MODE RAD
                                 ------------------------
(2) Store RPN programs
that take t from level 1         EXAMPLE: a=-2π ≤t≤ 2π=b
return X(t), Y(t), Z(t):          t → (cos(t),sin(t),t)

  « COS » ENTER 'X' STO          seen from (75,180,1456)
  « SIN » ENTER 'Y' STO
    «   » ENTER 'Z' STO          Proceed as with HP-28C&S

(3) Store a vector in W:

 [75 180 1456] 'W' STO

(4) Run curve:   curve
```

Exhibit 2. These programs (for the HP-28C and HP-28S on the left, for the HP-48SX on the right) trace on the screen the image of a curve in space. Notice that the length of each subroutine does not exceed about four lines, which enables the HP-28C to display each subroutine on its small screen.

points, a structured network of curves on the surface, called a *wire frame*, yields a more realistic picture of the surface, as shown in **Exhibit 3**. **Exhibit 4** lists the program.

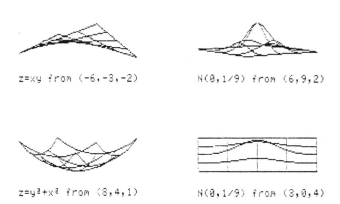

Exhibit 3. Sample of three-dimensional graphics produced with the HP-28C. The notation $N(\mu, \sigma^2)$ stands for the bivariate Gaussian distribution with mean vector μ and covariance matrix equal to σ^2 times the identity, $\sigma^2 I$.

To understand the use of wire frames in plotting functions of two variables, let $f : D \to \mathbb{R}$ represent a function defined on a rectangular domain,

$$D = [a, b] \times [c, d] = \{(x, y) \in \mathbb{R}^2 | a \leq x \leq b \text{ and } c \leq y \leq d\}.$$

(Non-rectangular domains accommodate wire frames as well, but in the present discussion, they would require additional abstraction without providing further enlightenment.) To construct a wire frame on f, put a grid on the domain D and then map the grid into (the graph of) f by means of the function f itself, as follows.

First select finitely many points, $a = x_1 < x_2 < \ldots < x_{m-1} < x_m = b$, the number of which, m, depends only upon the desired resolution and the allowed computational time. For each selected x_i, let

$$X_i = \{(x, y) \in D | x = x_i \text{ and } c \leq y \leq d\}$$

denote the straight line segment parallel to the y-axis and spanning the rectangle D at the abscissa x_i. (The segments X_i form one half of the grid.) Then map each line segment X_i into a curve γ_i on the surface f, according to the rule

$$\gamma_i : [c, d] \to \mathbb{R}^3, \quad t \mapsto \begin{pmatrix} x_i \\ t \\ f(x_i, t) \end{pmatrix};$$

3-D Graphics in Calculus and Linear Algebra   153

```
-------------------------
POINT IN 3D FOR HP-28C&S
-------------------------
point
« tests tips CLLCD ΣDAT
ARRY→ 3 DROPN dots R→C
PIXEL 1 3 START line
NEXT ( point ) ORDER »

tests
« DEPTH 0 == X IFT 'X'
STO W TYPE 3 == W 1 1 1
{ 3 } →ARRY IFTE 'W' STO
basis CLΣ »

basis
« 'W' W ABS INV STO* 0 0
1 { 3 } →ARRY DUP W DOT
W * - DUP ABS INV * 'V'
STO V W CROSS 'U' STO »

tips
« 0 0 0 X ARRY→ DROP 3
DUPN DROP 0 1 0 0 0 1 0
0 0 1 1 6 START map { 2
} →ARRY Σ+ NEXT true »

true
« SCLΣ PPAR LIST→ 4
DROPN - NEG C→R 32 /
SWAP 137 / DUP2 DUP2 IF
> THEN / DUP *W ELSE
SWAP / DUP *H END ROT
ROT MIN * 'side' STO »

dots
« many 'n' 4 STO/ ray »

line
« many ray »

many
« R→C 'b' STO b ABS side
/ 'n' STO »

ray
« 6 n / 'tiny' STO tiny
NEG 0 n START tiny + DUP
PIXEL NEXT DROP »

map
« { 3 } →ARRY DUP U DOT
SWAP V DOT »
-------------------------
EXAMPLE: X=(1,2,3) in 3D
 [ 1 2 3 ]  USER  point

-------------------------
X=(3,4,2) from W=(6,3,2)
   [ 6 3 2 ]  'W'  STO
   [ 3 4 2 ]  USER  point
```

```
-------------------------
A POINT IN 3D ON HP-48SX
-------------------------
point
« tests basis CLΣ tips
true ΣDAT lines { point
} ORDER GRAPH
»

tests
« DEPTH 0 == X IFT 'X'
STO W TYPE 3 == W 1 1 1
→V3 IFTE 'W' STO
»

basis
« 'W' W ABS INV STO* 0 0
1 →V3 DUP W DOT W * -
DUP ABS INV * 'V' STO V
W CROSS 'U' STO
»

tips
« 0 0 0 X V→ DROP 0 X V→
1 0 0 0 1 0 0 1 1 6
  START map →V2 Σ+
  NEXT
»

true
« SCATRPLOT MAXΣ MINΣ -
V→ 6.3 / SWAP 13 / / DUP
  IF 1 >
  THEN *W
  ELSE INV *H
  END ERASE
»

lines
« OBJ→ 3 DROPN R→C PIXON
R→C (0,0) TLINE 1 3
  START R→C (0,0) LINE
  NEXT
»

map
« →V3 DUP U DOT SWAP V
DOT
»
-------------------------
EXAMPLE: X=(1,2,3) in 3D
 [ 1 2 3 ]    VAR  point

-------------------------
X=(3,4,2) from W=(6,3,2)
   [ 6 3 2 ]  'W'  STO
   [ 3 4 2 ]  VAR  point
```

Exhibit 4. This program delineates the image on the screen of the graph (surface) of a function of two variables. Notice that spelling the name of each subroutine in lower-case letters differentiates them, especially in subsequent uses, from the HP-28C&S's or HP-48SX's own routines (the names of which always appear in capital letters).

and plot each curve γ_i as indicated in the preceding section, by means of finitely many points on it, $\gamma_i(t_1), \ldots, \gamma(t_N)$. Choose N much larger than m, so as to give the impression of clearly distinct (with m small) continuous (with N large) curves.

Similarly, select finitely many points, $c = y_1 < y_2 < \ldots < y_{n-1} < y_n = d$, and for each of them let

$$Y_j = \{(x,y) \in D | a \leq x \leq b \text{ and } y = y_j\}$$

denote the straight line segment parallel to the x-axis and spanning the rectangle D at the ordinate y_j. (The segments Y_j, together with the orthogonal segments X_i, form the grid on the rectangle D.) Then map each segment Y_j into a curve Γ_j on (the graph of) f, according to the rule

$$\Gamma_j : [a,b] \to \mathbb{R}^3, \qquad s \mapsto \begin{pmatrix} s \\ y_j \\ f(s, y_j) \end{pmatrix},$$

and plot the curve Γ_j by choosing finitely many points, $\Gamma_j(s_1), \ldots, \Gamma_j(s_M)$, with $a \leq s_1 < s_2 < \ldots < s_{M-1} < s_M \leq b$. (Again, choose M much larger than n.) Together, the curves γ_i and the curves Γ_j constitute the desired network of curves, or "wire frame," that outlines the graph of f, as demonstrated in **Exhibit 3**.

5. Rendering

The algorithm developed thus far produces pictures that suggest not an opaque surface but a wire-frame skeleton of a surface. In order to furnish a more realistic picture of an opaque surface, an algorithm must also omit all portions of the wire frame that remain hidden from the observer by other pieces of the opaque surface. The procedure that determines what portions of the wire frame to omit is called a *hidden-line remover;* the present section describes two different procedures to determine hidden lines.

Hidden-line removers require not only the direction of the observer, as do axonometries, but also the observer's location (which amounts to the direction *and* the distance from the origin). The observer's location serves to determine which parts of the surface lie in front of the observer and which parts lie hidden by other portions of the surface. For the purposes of the present section, it suffices to assume that the observer stands "away" from the only object in the picture (the graph of a function of two variables, $f : D = [a,b] \times [c,d] \to \mathbb{R}$). Here, the word "away" means that if the observer stands at $\vec{s} = (s_x, s_y, s_z)$, then $(s_x, s_y) \notin [a,b] \times [c,d]$: the observer may not stand directly above or below the domain of f.

To illustrate the effect of a hidden-line remover, **Exhibit 5** displays the graph of the function f defined by

$$f(x,y) = -\left(x^2 - 1\right)^2 - \left(x^2 - e^y\right)^2,$$

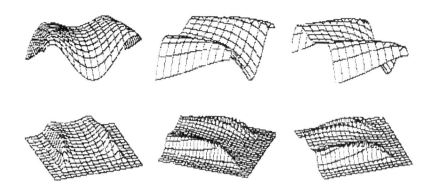

Exhibit 5. Views of the graph of $z = -(x^2 - 1)^2 - (x^2 - e^y)^2$. The top row shows the portion where $(x, y) \in [-3/2, 3/2] \times [-4/5, 4/5]$. The bottom row displays the entire graph, rescaled as $\zeta = \text{Arctan} f(\tan \xi, \tan \eta)$. The observer stands at $(2, 6, 9)$ for **(a)** and **(d)**, at $(6, 2, 9)$ for **(b)** and **(e)**, and at $(6, -2, 9)$ for **(c)** and **(f)**.

with hidden lines removed. This function was constructed by Bjorn Poonen [1987], then a student at Harvard College. It has two local maxima, at $(-1, 0)$ and $(1, 0)$, but no other critical point.

5.1 "Expanding Horizons" to Remove Hidden Lines

For detecting hidden objects and omitting them from the plot, the standard method of "expanding horizons" (a variant of the "variable horizon" outlined in Peitgen and Saupe [1988, 128], and also called the "perimeter method" in Plastock and Kalley [1986]) proceeds as follows.

Construct a matrix $H \in \mathbb{M}_{2 \times p}(\mathbb{R})$ with two rows—the top and bottom horizons—and p columns, with p equal to the number of pixels in the length of the screen. Initially, set $H(1, j) = -\infty$ and $H(2, j) = +\infty$ for every $j \in \{1, \ldots, p\}$ (on a computer, substitute the greatest machine number for ∞). The two horizons keep records of the highest and lowest points already plotted in the j-th column of pixels, thus indicating that new images coming between the two horizons must remain hidden. Next, while tracing the families of curves just described on f, plot the curves closest to the observer first, followed by those further and further away, subject to the following rule:

$$\begin{cases} \text{if } H(2, j) < v < H(1, j), \text{ then do } not \text{ plot } (u, v); \\ \text{if } v \leq H(2, j), \text{ then plot } (u, v) \text{ and reset } H(2, j) = v; \\ \text{if } H(1, j) \leq v, \text{ then plot } (u, v) \text{ and reset } H(1, j) = v. \end{cases}$$

This method served to plot the graphics in **Exhibits 4** and **5**.

5.2 "Ray Tracing" to Remove Hidden Lines and to Shade Surfaces

The method of expanding horizons succeeds quickly for 3-D graphics displaying one function of two variables, but the version described in the preceding subsection may fail for other types of surfaces, such as spheres, tori, or collections of more than one function, which require a more versatile hidden-line remover, for instance, the "ray tracing" procedure described here.

A *ray tracing* procedure omits hidden lines and hidden points according to the following principle: a point $\vec{p}$ remains hidden from the observer at $\vec{s}$ if, but only if, there exists an object between the hidden point and the observer. Consequently, to determine whether the observer at $\vec{s}$ can or cannot see the point $\vec{p}$, it suffices to determine whether or not any of the objects in the picture intersects the straight line segment from $\vec{p}$ through $\vec{s}$, denoted by $[\vec{p}, \vec{s}]$, at a point other than $\vec{p}$.

Specifically, imagine a picture showing only one surface, denoted by Σ (for instance, a sphere, a torus, or an ellipsoid), and assume that the surface Σ consists of the points (x, y, z) that satisfy some given equation $F(x, y, z) = 0$. (Technically, F stands for a continuously differentiable function $F : \mathbb{R}^3 \to \mathbb{R}$ such that its gradient vanishes nowhere on the surface: $\overrightarrow{\text{grad}}\, F(x, y, z) \neq \vec{0}$ at each point (x, y, z) where $F(x, y, z) = 0$.) Moreover, consider any point $\vec{p}$ on Σ. To determine whether the observer at $\vec{s}$ can see the point $\vec{p}$, first construct a parametric representation of the segment $[\vec{p}, \vec{s}]$ from $\vec{p}$ through $\vec{s}$, for example by means of the function

$$\gamma : [0, 1] \to \mathbb{R}^3, \qquad t \mapsto (1 - t)\vec{p} + t\vec{s} = \vec{p} + t(\vec{s} - \vec{p}).$$

Second, observe that the surface Σ has two sides, in effect one side where $F(x, y, z) < 0$ and one side where $F(x, y, z) > 0$. Thus, the segment $[\vec{p}, \vec{s}]$ "pierces" the surface Σ at some point $\gamma(t_0) \in \Sigma$ if, but only if, the composite function $F \circ \gamma$ changes sign at some value $t_0 \in [0, 1]$. (At the expense of greater detail, the user need only consider the straight line segment from $\vec{p}$ through the object closest to $\vec{s}$ on $[\vec{p}, \vec{s}]$, instead of the entire segment from $\vec{p}$ through $\vec{s}$.)

This means that it suffices to program just one function of a single variable, $\phi : [0, 1] \to \mathbb{R}$, $\approx \mapsto (F \circ \gamma)(\approx)$, and to apply a "coarse" equation solver to determine if ϕ changes sign inside the interval $]0, 1[= \{t : 0 < t < 1\}$. The equation solver need only be "coarse" in the sense that it need not estimate *where* but only *whether* ϕ might change sign.

Programming Exercise

27. Consider a three-dimensional graphic displaying one object, in effect the graph (or surface) of a function of two variables, seen by an observer at a site $\vec{s}$, and lit by a light source from a direction $\vec{\ell}$. (Assume parallel light

beams from the direction $\vec{\ell}$, to avoid the need for projective geometry.) For an example, see **Exhibit 6**.

a) Design a strategy, and describe it in prose, for determining whether a point $\vec{p}$ lies in the shadow of the surface (as opposed to in the light from $\vec{\ell}$).

b) Write a routine that implements the strategy described in the preceding item, and test it.

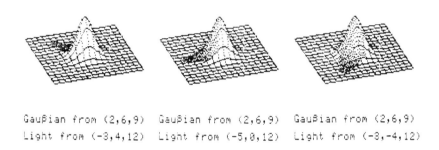

Gaußian from (2,6,9)   Gaußian from (2,6,9)   Gaußian from (2,6,9)
Light from (-3,4,12)   Light from (-5,0,12)   Light from (-3,-4,12)

Exhibit 6. 3-D graphics with hidden lines and shading, produced by the HP-28S, with two screens for each graphic.

6. Term Projects

6.1 From Computer Programs to Number Theory

This project demonstrates how even such inelegant practical problems as that of testing a computer program may lead to interesting problems in abstract mathematics, for instance, in number theory. To this effect, this project addresses the following two questions:

- Are there easier numerical examples than those presented in the text?

- Are there test examples for which floating point computations occur without errors?

The answers to both questions depends upon whether orthonormal bases $(\vec{u}, \vec{v}, \vec{w})$ for $\mathbb{R}^3$ exist with only rational coordinates having finite decimal or binary expansions. Such bases would never demand the extraction of square roots and would not require infinitely many digits. Thus, they would lend themselves better than would other bases to hand calculations and to error-free computations in floating point arithmetic.

To answer both questions, you may proceed as follows.

Task #1. Read an introduction to Pythagorean triples (triples of integers (h, k, ℓ) such that $h^2 + k^2 = \ell^2$), e.g., Nishi [1987] or van der Waerden [1983, 1–8].

Task #2. Read the article by Osborne and Liebeck [1989] and extract from it a numerical recipe for building examples of orthonormal bases $(\vec{u}, \vec{v}, \vec{w})$ with only rational coordinates *and* such that $u_z = 0$. (Recall that the vector $\vec{u}$ must appear horizontal on the screen, which means that its vertical coordinate must vanish.)

Task #3. Determine the "simplest" orthonormal basis $(\vec{u}, \vec{v}, \vec{w})$ for which each coordinate has a finite decimal expansion, but in which none of the basis vectors coincides with any of the directions in the canonical basis. (Do not use $(\pm 1, 0, 0), (0, \pm 1, 0)$, or $(0, 0, \pm 1)$, as they do not lead to visually representational axonometries.)

Task #4. Find *all* orthonormal bases $(\vec{u}, \vec{v}, \vec{w})$ as just described for which each coordinate has a finite *binary* expansion.

6.2 Perspectives without Projective Geometry

Recall that an axonometry projects the three-dimensional space orthogonally onto the screen W, along a direction parallel to a prescribed vector $\vec{w}$ perpendicular to the screen; thus, an axonometry is an orthogonal projection.

In contrast, a *perspective* is a *central projection* from a prescribed point C outside the screen. The point C is called the *center* and it represents the eye of the observer. Specifically, a perspective is a function

$$P : \mathbb{R}^3 \setminus \{C\} \to W,$$

for which the image $P(\vec{x})$ of each point $\vec{x} \neq C$ in space lies at the intersection of the screen and the straight line through C and $\vec{x}$.

Assume that the center C lies at distance c from the origin in the direction $\vec{w}$; thus, $C = c\vec{w}$. Modify **Algorithm 2** so that for each point $\vec{x}$ in space the algorithm determines the intersection of the screen and the straight line through $\vec{x}$ and $c\vec{w}$. The new algorithm should differ from **Algorithm 2** only by a scaling factor, which depends upon C and $\vec{x}$.

6.3 Geometric Perspectives on History

Write a paper about the history of the role of geometry in 3-D graphics. Besides prose, include specific theories and detailed examples, and cover at least the following three topics:

- Agatharchos's uses of perspectives to design the stage for Aechylus's tragedies about A.D. 450. You may start with van der Waerden's article [1954].

- Geometric and photographic evidence of Robert E. Peary's polar expeditions, described in Davies's article [1990a] and report [1990b].

- 3-D computer graphics. See Hanes [1991] and Greenberg [1989].

6.4 2-D Graphics in Flatland

Write a paper, with theorems, algorithms, computer programs, and examples of 2-D graphics, with line-removers and shading, for the inhabitants of "Flatland," which is a two-dimensional universe with one-dimensional screens. You may, but need not, read Abbott [1891].

7. Solutions to the Exercises

1. $\langle (2,3,6), (8,1,4) \rangle = 2 \cdot 8 + 3 \cdot 1 + 6 \cdot 4 = 16 + 3 + 24 = 43$.

2. $\|(2,3,6)\| = \sqrt{2^2 + 3^2 + 6^2} = \sqrt{4 + 9 + 36} = \sqrt{49} = 7$.

3. Verify that the following three dot products equal zero:

   $\langle \vec{u}, \vec{v} \rangle = \langle (2,3,6), (3,-6,2) \rangle = (2 \times 3) + (3 \times (-6)) + (6 \times 2) = 6 - 18 + 12 = 0$,

   $\langle \vec{v}, \vec{w} \rangle = \langle (3,-6,2), (6,2,-3) \rangle = (3 \times 6) + (-6 \times 2) + (2 \times (-3)) = 18 - 12 - 6 = 0$,

   $\langle \vec{w}, \vec{u}, \rangle = \langle (6,2,-3), (2,3,6) \rangle = (6 \times 2) + (2 \times 3) + (-3 \times 6) = 12 + 6 - 18 = 0$.

   By symmetry of the dot product, $\langle \vec{v}, \vec{u} \rangle = \langle \vec{u}, \vec{v} \rangle = 0$, $\langle \vec{w}, \vec{v} \rangle = \langle \vec{v}, \vec{w} \rangle = 0$, and $\langle \vec{u}, \vec{w} \rangle = \langle \vec{w}, \vec{u} \rangle = 0$.

4. First, verify that they each have length one: $\|\vec{u}\| = \sqrt{\langle \vec{u}, \vec{u} \rangle} = \sqrt{4/49 + 9/49 + 36/49} = \sqrt{49/49} = 1$. Similarly, $\|\vec{v}\| = 1 = \|\vec{w}\|$. Second, verify that they are mutually orthogonal: $\langle \vec{u}, \vec{v} \rangle = 2/7 \cdot 3/7 - 3/7 \cdot 6/7 + 6/7 \cdot 2/7 = 0$. Similarly, $\langle \vec{v}, \vec{w} \rangle = 0$ and $\langle \vec{w}, \vec{u} \rangle = 0$.

5. $a = \langle \vec{x}, \vec{u} \rangle = \langle (1,2,3), (2/7, 3/7, 6/7) \rangle = 1 \cdot 2/7 + 2 \cdot 3/7 + 3 \cdot 6/7 = 26/7$. Similarly, $b = \langle \vec{x}, \vec{v} \rangle = -3/7$ and $c = \langle \vec{x}, \vec{w} \rangle = 1/7$. As a verification,

   $$26/7(2/7, 3/7, 6/7) - 3/7(3/7, -6/7, 2/7) + 1/7(6/7, 2/7, -3/7) = (1,2,3).$$

6. $P(\vec{x}) = a\vec{u} + b\vec{v} = 26/7(2/7, 3/7, 6/7) - 3/7(3/7, -6/7, 2/7) = (43/49, 96/49, 150/49)$.

7. $(4,1,8) \times (4,-8,-1) = (63, 36, -36) = 9(7, 4, -4)$.

8. $\|\vec{h}\| = \|\vec{k}\| = \|\vec{\ell}\| = 1$ and $\langle \vec{h}, \vec{k} \rangle = \langle \vec{k}, \vec{\ell} \rangle = \langle \vec{\ell}, \vec{h} \rangle = 0$.

9. $a = \langle \vec{x}, \vec{h} \rangle = 10/3$, $b = \langle \vec{x}, \vec{k} \rangle = -5/3$, and $c = \langle \vec{x}, \vec{\ell} \rangle = 1/3$. As a verification, $a\vec{h} + b\vec{k} + c\vec{\ell} = (1,2,3)$.

10. $P(\vec{x}) = a\vec{h} + b\vec{k} = (20/27, 50/27, 85/27)$.

11. First, verify that $\|(2/11, 6/11, 9/11)\| = 1 = \|(9/11, -6/11, 2/11)\|$ and that $\langle (2/11, 6/11, 9/11), (9/11, -6/11, 2/11) \rangle = 0$. Second, let

    $$\vec{t} = \vec{r} \times \vec{s} = (2/11, 6/11, 9/11) \times (9/11, -6/11, 2/11) = (6/11, 7/11, -6/11).$$

    Finally, verify that $\|\vec{t}\| = \|(6/11, 7/11, -6/11)\| = 1$.

12. The following calculations verify that $\vec{x}$ is perpendicular to $\vec{x} \times \vec{y}$:

$$\langle (x_1, x_2, x_3), (x_1, x_2, x_3) \times (y_1, y_2, y_3) \rangle$$
$$= \langle (x_1, x_2, x_3), (x_2 y_3 - x_3 y_2, x_3 y_1 - x_1 y_3, x_1 y_2 - x_2 y_1) \rangle$$
$$= x_1(x_2 y_3 - x_3 y_2) + x_2(x_3 y_1 - x_1 y_3) + x_3(x_1 y_2 - x_2 y_1)$$
$$= x_1 x_2 y_3 - x_1 x_3 y_2 + x_2 x_3 y_1 - x_2 x_1 y_3 + x_3 x_1 y_2 - x_3 x_2 y_1 = 0.$$

Similar calculations verify that $\vec{y}$ is perpendicular to $\vec{x} \times \vec{y}$.

13. The constraint that $(x, y, z) = \vec{x} \in W$ means that $\langle \vec{x}, \vec{w} \rangle = 0$, because by definition the plane W consists of all vectors perpendicular to $\vec{w}$. Consequently, $\vec{x} \in W$ if, but only if, $\langle (x, y, z), (m, r, s) \rangle = 0$, which means that (x, y, z) belongs to the zero set of the constraint function

$$g : \mathbb{R}^3 \to \mathbb{R}, \quad g(x, y, z) = \langle (x, y, z), (m, r, s) \rangle = xm + yr + zs.$$

To minimize the function $f : \mathbb{R}^3 \to \mathbb{R}$ defined by

$$f(x, y, z) = d(\vec{x_0}, \vec{x})^2 = (x - x_0)^2 + (y - y_0)^2 + (z - z_0)^2$$

subject to the constraint that

$$0 = (x, y, z) = xm + yr + zs,$$

write the Lagrangian function

$$F(x, y, z, \ell) = f(x, y, z) - \ell \cdot g(x, y, z)$$
$$= (x - x_0)^2 + (y - y_0)^2 + (z - z_0)^2 - \ell \cdot (xm + yr + zs).$$

Next, set the gradient (all partial derivatives) of F equal to zero; with D_i denoting partial differentiation with respect to the ith variable, this gives

$$\begin{cases} D_1 F(x, y, z, \ell) = 0: & 2(x - x_0) - \ell m = 0, \\ D_2 F(x, y, z, \ell) = 0: & 2(y - y_0) - \ell r = 0, \\ D_3 F(x, y, z, \ell) = 0: & 2(z - z_0) - \ell s = 0, \\ D_4 F(x, y, z, \ell) = 0: & -(xm + yr + zs) = 0. \end{cases}$$

If none of the coordinates m, r, and s equals zero, eliminating Lagrange's multiplier ℓ from the first three equations yields

$$\ell = \frac{2(x - x_0)}{m} = \frac{2(y - y_0)}{r} = \frac{2(z - z_0)}{s}.$$

The equations just established correspond to the straight line through $\vec{x_0} = (x_0, y_0, z_0)$ and parallel to $\vec{w} = (m, r, s)$, which is hence perpendicular to W. (The same conclusion holds in each of the particular cases where one or two, but not all three, of $m = 0$, or $r = 0$, or $s = 0$, since $\vec{w} = (m, r, s) \neq \vec{0}$.) The fourth equation, $(-xm + yr + zs) = 0$, still means that the solution, (x, y, z), lies in the plane W. Therefore, the method of Lagrange's multiplier reveals that *the point $\vec{x}$ in W that lies closest to $\vec{x_0}$ is at the intersection of W and the straight line through $\vec{x_0}$ perpendicular to W*.

14. $\vec{w} = (9/25, 12/25, 20/25) = (0.36, 0.48, 0.80).$
$\vec{v} = (-12/25, -16/25, 15/25) = (-0.48, -0.64, 0.60).$
$\vec{u} = (-4/5, 3/5, 0) = (-0.8, 0.6, 0.0).$

15. $u = 2/5 = 0.40$ and $v = 1/25 = 0.04.$

16. $u = \langle \vec{t}, \vec{u} \rangle = xu_x + yu_y + zu_z$ and $v = \langle \vec{t}, \vec{v} \rangle = xv_x + yv_y + zv_z.$

17. First, calculate $\|\vec{s}\| = \|(75, 180, 1456)\| = 1469 = 13 \times 113.$ Hence,

$$\vec{w} = \frac{1}{\|\vec{s}\|} \vec{s} = \frac{1}{1469}(75, 180, 1456) = (75/1469, 180/1469, 1456/1469).$$

Second, calculate

$$P(\vec{z}) = \vec{z} - \langle \vec{z}, \vec{w} \rangle \vec{w} = (0, 0, 1) - \frac{1456}{1469^2}(75, 180, 1456) = \ldots$$

$$= \frac{1}{1469^2}(-109{,}200, -262{,}080, 38{,}025) = \frac{15}{1469^2}(-7280, -17{,}472, 2535).$$

Third, calculate $\|P(\vec{z})\| = (15/1469^2) \times 19{,}097$ and

$$\vec{v} = \frac{1}{\|P(\vec{z})\|} P(\vec{z}) = \ldots = \frac{1}{1469}(-560, -1344, 195).$$

Finally,

$$\vec{u} = \vec{v} \times \vec{w} = \frac{1}{1469^2}(1{,}991{,}964, -829{,}985, 0) = \ldots = (12/13, -5/13, 0).$$

Next, to determine the image (u, v) of $\vec{r} = (1, 2, 3)$, calculate the dot products

$$u = \langle \vec{r}, \vec{u} \rangle = \langle (1, 2, 3), (12/13, -5/13, 0) \rangle = 2/13 = 0.153\,846\,153\,846\ldots,$$

$$v = \langle \vec{r}, \vec{v} \rangle = \langle (1, 2, 3), (-560/1469, -1344/1469, 195/1469) \rangle$$
$$= -2663/1469 = -1.812\,797\,821\ldots;$$

thus, the point $(1, 2, 3)$ appears at $(u, v) \approx (0.154, -1.813)$ on the screen.

18. First, observe that $K := (0, 0, 1) \times H$ is perpendicular to the vertical vector $(0, 0, 1) = \vec{e}_3$; hence, K is horizontal. Moreover, K is perpendicular to H. Consequently, K is horizontal and perpendicular to H.

Second, observe that $L := H \times K$ is perpendicular to H; hence, K and L both lie in the plane perpendicular to H. Furthermore, L is also perpendicular to K, which is horizontal.

Therefore, (H, K, L) forms an orthogonal system, with H pointing perpendicularly to a plane containing the horizontal vector K and the perpendicular vector L.

Thus, if none of H, K, and L is the zero vector, then the corresponding unit vectors ($H/\|H\|$, $K/\|K\|$, and $L/\|L\|$) form an orthonormal basis pointing along the three perpendicular directions for 3-D graphics from the direction of view along H. Then the present algorithm constitutes an alternative method for computing the basis $(\vec{u}, \vec{v}, \vec{w}) = (K/\|K\|, L/\|L\|, H/\|H\|)$.

19. To construct the matrix of $i \circ P$ relative to the bases $(\vec{u}, \vec{v}, \vec{w})$ and $(\vec{u}, \vec{v})$, recall that the columns of the matrix must be the images (by $i \circ P$) of the first basis vectors, expressed in terms of the second basis. In this particular case, since P restricts to the identity on its range (spanned by $\vec{u}$ and $\vec{v}$),
$$P(\vec{u}) = \vec{u} = 1 \cdot \vec{u} + 0 \cdot \vec{v}$$
and the first column of the matrix has entries 1 and 0. Similarly,
$$P(\vec{v}) = \vec{v} = 0 \cdot \vec{u} + 1 \cdot \vec{v}$$
and the second column has entries 0 and 1. Finally,
$$P(\vec{w}) = \vec{0} = 0 \cdot \vec{u} + 0 \cdot \vec{v}$$
and thus the third column has entries 0 and 0. Consequently,
$$A = \begin{pmatrix} 1 & 0 & 0 \\ 0 & 1 & 0 \end{pmatrix}.$$

20. To construct the matrix of P relative to the canonical basis $(\vec{x}, \vec{y}, \vec{z})$ for $\mathbb{R}^3$ and the basis $(\vec{u}, \vec{v})$ for the screen, utilize the orthonormality of both bases.
$$P(\vec{x}) = \langle \vec{x}, \vec{u} \rangle \vec{u} + \langle \vec{x}, \vec{v} \rangle \vec{v} = u_x \vec{u} + v_x \vec{v},$$
which reveals that the first column has entries u_x and v_x. Similarly,
$$P(\vec{y}) = \langle \vec{y}, \vec{u} \rangle \vec{u} + \langle \vec{y}, \vec{v} \rangle \vec{v} = u_y \vec{u} + v_y \vec{v},$$
and the second column has entries u_y and v_y. Finally,
$$P(\vec{z}) = \langle \vec{z}, \vec{u} \rangle \vec{u} + \langle \vec{z}, \vec{v} \rangle \vec{v} = u_z \vec{u} + v_z \vec{v};$$
thus, the third column has entries u_z and v_z, which gives the matrix
$$B = \begin{pmatrix} u_x & u_y & u_z \\ v_x & v_y & v_z \end{pmatrix} = \begin{pmatrix} \vec{u} \text{ in terms of the canonical basis} \\ \vec{v} \text{ in terms of the canonical basis} \end{pmatrix}.$$
For example, with the particular vectors given in the exercise,
$$B = \begin{pmatrix} u_x & u_y & u_z \\ v_x & v_y & v_z \end{pmatrix} = \begin{pmatrix} 4/5 & 3/5 & 0 \\ -36/65 & -48/65 & 25/65 \end{pmatrix}.$$

21. First, write the matrix of $i \circ P$ relative to the bases $(\vec{x}, \vec{y}, \vec{z})$ and $(\vec{u}, \vec{v}, \vec{w})$. The calculations in the previous exercise showed that the images $P(\vec{x}), P(\vec{y})$, and $P(\vec{z})$ do not involve $\vec{w}$. Therefore, it suffices to append a row of zeroes to the matrix just constructed in order to reflect these components, which produces

$$C = \begin{pmatrix} u_x & u_y & u_z \\ v_x & v_y & v_z \\ 0 & 0 & 0 \end{pmatrix}.$$

Second, construct the matrix for the change of bases from the basis $(\vec{u}, \vec{v}, \vec{w})$ to the canonical basis $(\vec{x}, \vec{y}, \vec{z})$. This matrix must transform the components of a vector in terms of the first basis into the coordinates of the same vector relative to the canonical basis. Consequently, this matrix has columns $\vec{u}, \vec{v}$, and $\vec{w}$ (in terms of the canonical basis):

$$D = (\vec{u} \ \vec{v} \ \vec{w}) = \begin{pmatrix} u_x & v_x & w_x \\ u_y & v_y & w_y \\ u_z & v_z & w_z \end{pmatrix}.$$

Finally, the matrix of $i \circ P$ with respect to the canonical basis in both the domain and the range equals the product of D and C,

$$E = D \cdot C = \begin{pmatrix} u_x & v_x & w_x \\ u_y & v_y & w_y \\ u_z & v_z & w_z \end{pmatrix} \cdot \begin{pmatrix} u_x & u_y & u_z \\ v_x & v_y & v_z \\ 0 & 0 & 0 \end{pmatrix}$$

$$= \begin{pmatrix} u_x^2 + v_x^2 & u_x u_y + v_x v_y & u_x u_z + v_x v_z \\ u_y u_x + v_y v_x & u_y^2 + v_y^2 & u_y u_z + v_y v_z \\ u_z u_x + v_z v_x & u_z u_y + v_z v_y & u_z^2 + v_z^2 \end{pmatrix}.$$

For example, with the particular vectors in the previous exercise,

$$E = \begin{pmatrix} -4/5 & -36/65 & 3/13 \\ 3/5 & -48/65 & 4/13 \\ 0 & 25/65 & 12/13 \end{pmatrix} \cdot \begin{pmatrix} -4/5 & 3/5 & 0 \\ -36/65 & -48/65 & 25/65 \\ 0 & 0 & 0 \end{pmatrix}$$

$$= \frac{1}{169} \begin{pmatrix} 160 & -12 & -36 \\ -12 & 153 & -48 \\ -36 & -48 & 25 \end{pmatrix}.$$

22. (A) If $\vec{s} = 0\vec{z} = \vec{0}$, then the projection is undefined. (B) If $\vec{s} = c\vec{z}$ with $c \neq 0$, then $\vec{z} = (1/c)\vec{s} \in \text{Kernel}(P)$; hence, $P(\vec{z}) = \vec{0}$ cannot appear vertical on the screen. By contraposition, the condition that $\vec{v}$ must appear vertical implies that $\vec{s}$ must be chosen so that $P(\vec{z}) \neq \vec{0}$. (C) Sketches of functions of two variables would appear as flat grids on the screen, without revealing any feature. (D) Step (3) in **Algorithm 1** would fail, because of the attempt to divide by $\|P(\vec{z})\|$, which equals zero.

3-D Graphics in Calculus and Linear Algebra   165

```
------------------------
WIRE FRAMES FOR HP-28C&S
------------------------
mesh
« basis fit unit steps
CLLCD bars rods { mesh }
ORDER PRLCD »

basis
« 'W' W ABS INV STO* 0 0
1 { 3 } →ARRY DUP W DOT
W * - DUP ABS INV * 'V'
STO V W CROSS 'U' STO »

fit
« -1 1 FOR x -1 1 FOR y
x y DUP2 Z map { 2 }
→ARRY Σ+ .5 STEP .5 STEP
SCLΣ »

unit
« MAXΣ MINΣ CLΣ - ARRY→
DROP2 137 / 'du' STO »

steps
« du U ARRY→ DROP2 V
ARRY→ DROP2 ROT R→C ABS
du SWAP / 'dy' STO R→C
ABS / 'dx' STO »

bars
« -1 1 FOR y -1 1 FOR x
x y DUP2 Z map R→C PIXEL
dx STEP .5 STEP »

rods
« -1 1 FOR x -1 1 FOR y
x y DUP2 Z map R→C PIXEL
dy STEP .5 STEP »

map
« { 3 } →ARRY DUP U DOT
SWAP V DOT »
------------------------
  EXAMPLE: z = y² - x²
  seen from  (12,9,20)

(1) Store a vector in W:

  [12 9 20] ENTER 'W' STO

(2) Put RPN program in Z
  that takes x in level 2,
  y in level 1, returns z:

  « SQ SWAP SQ - » ENTER
           'Z' STO

(3) Run mesh:  USER mesh
```

```
------------------------
WIRE FRAMES FOR  HP-48SX
------------------------
mesh
« basis fit unit steps
bars rods { mesh } ORDER
GRAPH
»

basis
« 'W' W ABS INV STO* 0 0
1 →V3 DUP W DOT W * -
DUP ABS INV * 'V' STO V
W CROSS 'U' STO
»

fit
« -1 1
  FOR x -1 1
    FOR y x y DUP2 Z map
→V2 Σ+ .2
    STEP .2
  STEP SCATRPLOT ERASE
»

unit
« MAXΣ MINΣ CLΣ - V→
DROP 131 / 'du' STO
»

steps
« du U V→ DROP V V→ DROP
ROT R→C ABS du SWAP /
'dy' STO R→C ABS / 'dx'
STO
»

bars
« -1 1
  FOR y -1 1
    FOR x x y DUP2 Z map
R→C PIXON dx
    STEP .5
  STEP
»

rods
« -1 1
  FOR x -1 1
    FOR y x y DUP2 Z map
R→C PIXON dy
    STEP .5
  STEP
»

map
« →V3 DUP U DOT SWAP V
DOT »
------------------------
```

Exhibit 7. Programs to plot the image of the canonical basis and one point, on the HP-28C&S and HP-48SX.

23. See the routine "basis" in **Exhibit 7**.

24. See the routine "map" in **Exhibit 7**.

25. See **Exhibit 7**.

26. Recall that
$$P(\vec{t}) = \langle \vec{t}, \vec{u} \rangle \vec{u} + \langle \vec{t}, \vec{v} \rangle \vec{v}$$
for each $\vec{t} \in \mathbb{R}^3$, and apply this expression to $\vec{t} = \gamma(t)$. This produces
$$p_u(t) = \langle \gamma(t), \vec{u} \rangle = u_x \gamma_x(t) + u_y \gamma_y(t) + u_z \gamma_z(t),$$
$$p_v(t) = \langle \gamma(t), \vec{v} \rangle = v_x \gamma_x(t) + v_y \gamma_y(t) + v_z \gamma_z(t).$$

27. First, observe that a point $\vec{p}$ lies in the shadow if, but only if, either of the following two conditions holds:

(A) The segment from the point $\vec{p}$ to the light source at $\vec{\ell}$ (for a central source), or the half-line from the point $\vec{p}$ in the direction of the light source (for parallel light beams from a source at infinity in the direction of $\vec{\ell}$), intersects the surface between the point $\vec{p}$ and the light source.

In practice, the method of ray tracing may determine whether such an intersection exists.

(B) The light source and the observer are on opposite sides of the tangent plane to the surface at the point $\vec{p}$.

To test for this condition, use *any* nonzero vector $\vec{n}$ perpendicular to the surface at the point $\vec{p}$, and determine whether the scalar products $\langle \vec{s} - \vec{p}, \vec{n} \rangle$ and $\langle \vec{\ell} - \vec{p}, \vec{n} \rangle$ have opposite signs: the sign of $\langle \vec{s} - \vec{p}, \vec{n} \rangle$ tells the side of the surface where the observer stands, whereas the sign of $\langle \vec{\ell} - \vec{p}, \vec{n} \rangle$ tells the side of the surface from which the light comes.

In practice, approximate the normal vector $\vec{n}$ by using two neighbors of $\vec{p}$ lying in two different directions on the surface, $\vec{q}$ and $\vec{r}$, and calculating the cross product $\vec{n} \approx (\vec{q} - \vec{p}) \times (\vec{r} - \vec{p})$.

Second, notice that points hidden from the observer must remain hidden, hence, unmarked on the screen, even if they lie in the shadow, which gives rise to a third rule.

(C) Shade (mark on the screen) a point in the shadow only if the observer can see that point.

Conclusion. Shade a point $\vec{p}$ on the surface if, but only if, the point $\vec{p}$ satisfies the condition (A OR B) AND C.

References

3-D Graphics

Francis, George K. 1987. *A Topological Picturebook*. New York: Springer-Verlag.

Greenberg, Donald P. 1989. Light reflection models for computer graphics. *Science* 244 (14 April 1989): 166–173.

Peitgen, Heinz-Otto, and Dietmar Saupe, eds. 1988. *The Science of Fractal Images*. New York: Springer-Verlag.

Plastock, Roy A., and Gordon Kalley. 1986. *Theory and Problems of Computer Graphics*. New York: McGraw-Hill.

Geometry and Algebra

Abbott, E.A. 1991 [1891]. *Flatland: A Romance of Many Dimensions*. 2nd ed. Princeton, NJ: Princeton University Press.

Hanes, Kit. 1991. An Introduction to Analytical Projective Geometry and Its Applications. UMAP Modules in Undergraduate Mathematics and Its Applications: Module 710. Reprinted in *UMAP Modules: Tools for Teaching 1990*, edited by Paul J. Campbell, 111–150. Arlington, MA: COMAP, 1991.

Malkevitch, Joseph. 1991. *Geometry's Future*. Arlington, MA: COMAP.

Nishi, Akihiso. 1987. A method for obtaining Pythagorean triples. *American Mathematical Monthly* 94 (9): 869–871.

Osborne, Anthony, and Hans Liebeck. 1989. Orthogonal bases of $\mathbb{R}^3$ with integer coordinates and integer lengths. *American Mathematical Monthly* 96 (1): 49–53.

van der Waerden, Bartel Leenert. 1983. *Geometry and Algebra in Ancient Civilizations*. New York: Springer-Verlag.

Weil, André. 1987. *Number Theory: An Approach through History From Hammurapi To Legendre*. Boston, MA: Birkhäuser.

Geometric Perspectives on History

Davies, Thomas D. 1990a. New evidence places Perry at the Pole. *National Geographic* 177 (1): 44–61.

_____. 1990b. *Robert E. Perry at the North Pole*. Rockville, MD: Foundation for the Promotion of the Art of Navigation (Box 1126, Rockville, MD 20850).

van der Waerden, Bartel Leenert. 1954. Les mathématiques appliquées dans l'antiquité. *L'enseignement mathématique* IIe série, 1 (1): 44–55.

Uses of Computer Graphics and Visual Intuition in Calculus

Ash, J. Marshall, and Harlan Sexton. 1985. A surface with one local minimum. *Mathematics Magazine* 58 (3): 147–149.

Curjel, Caspar R. 1990. *Exercises in Multivariable and Vector Calculus.* New York: McGraw-Hill.

Poonen, Bjorn. 1987. Solution to Problem 1235. *Mathematics Magazine* 60 (1): 48.

Rosenholtz, Ira, and Lowell Smylie. 1985. The only critical point in town. *Mathematics Magazine* 58 (3): 149–150.

Thomas, Banchoff, and Student Associates. 1989. Student generated interactive software for calculus of surfaces in a workstation laboratory. *UME Trends* 1 (3): 8, 7.

Linear Algebra Texts with a Section on Computer Graphics

Gerber, Harvey. 1990. *Elementary Linear Algebra.* Pacific Grove, CA: Brooks/Cole.

McCann, Roger C. 1984. *Introduction to Linear Algebra.* San Diego, CA: Harcourt Brace Jovanovich.

Acknowledgments

The Wadsworth Publishing Company provided financial support for the preparation of the first draft of the present material.

Moreover, Herb Holden at Gonzaga University gave me an introduction, delivered while jogging, to the basis of computer graphics, and Randy S. Kimmerly, then a senior at Eastern Washington University, demonstrated the portability of the present theory by programming it on an IBM personal computer.

Furthermore, Dr. Stephen P. Keeler, Ph.D., mathematician and manager of the Geometry and Optimization group at Boeing Computer Services, Joyce de Vries Kehoe, writer in Seattle, and Gail Nord at Gonzaga University, kindly provided their editorial assistance during the preparation of the manuscript.

Finally, a generous Seed Grant (Grant Number 143150-92-02) from the Washington Center for Improving the Quality of Undergraduate Education, at Evergreen State College, in Olympia, WA, allowed for testing the penultimate version of the manuscript in a workshop with Washington State Community College mathematics instructors.

About the Author

Yves Nievergelt graduated in mathematics from the École Polytechnique Fédérale de Lausanne (Switzerland) in 1976, with concentrations in functional and numerical analysis of PDEs. He obtained a Ph.D. from the University of Washington in 1984, with a dissertation in several complex variables under the guidance of James R. King. He now teaches complex and numerical analysis at Eastern Washington University.

Prof. Nievergelt is an associate editor of *The UMAP Journal*. He is the author of several UMAP Modules, a bibliography of case studies of applications of lower-division mathematics (*The UMAP Journal* 6(2)(1985): 37–56), and *Mathematics in Business Administration* (Irwin, 1989).

A New Release From COMAP's Video Applications Library

VIDEO APPLICATIONS LIBRARY

Statistics: Decisions Through Data is an introductory statistics course that unravels the statistical arguments behind surveys, polls, experiments, and product claims.

The series is composed of five hour-long instructional videos that contain a total of twenty-one individual segments. Each segment includes documentary material, interviews, and worked problems using actual data. The series also includes a 250-page User's Guide, which includes video summaries, learning objectives, exercises, group activities, quizzes, and worked solutions.

Each episode begins with a documentary segment that engages students' interest, and then teaches skills to gather data, analyze patterns, and draw conclusions about real-world issues.

HOUR ONE: UNITS 1–6
BASIC DATA ANALYSIS

- WHAT IS STATISTICS?
- STEMPLOTS
- HISTOGRAMS AND DISTRIBUTIONS
- MEASURES OF CENTER
- BOXPLOTS
- THE STANDARD DEVIATION

#VL8001 $99

HOUR TWO: UNITS 7–10
DATA ANALYSIS FOR ONE VARIABLE

- NORMAL CURVES
- NORMAL CALCULATIONS
- STRAIGHT-LINE GROWTH
- EXPONENTIAL GROWTH

#VL8002 $99

HOUR THREE: UNITS 11–14
DATA ANALYSIS FOR TWO VARIABLES

- SCATTERPLOTS
- FITTING LINES TO DATA
- CORRELATION
- SAVE THE BAY

#VL8003 $99

HOUR FOUR: UNITS 15–18
PLANNING DATA COLLECTION

- DESIGNING EXPERIMENTS
- THE QUESTION OF CAUSATION
- CENSUS AND SAMPLING
- SAMPLE SURVEYS

#VL8004 $99

HOUR FIVE: UNITS 19–21
INTRODUCTION TO INFERENCE

- SAMPLING DISTRIBUTIONS
- CONFIDENCE INTERVALS
- TESTS OF SIGNIFICANCE

#VL8005 $99

ALL FIVE VIDEOS WITH TEXT #VL8006 $350   USER'S GUIDE ONLY #VL8007 $18

FOR MORE INFORMATION, OR TO ORDER, CALL TOLL-FREE
1–800–77–COMAP
(1–800–772–6627).

VAL

UMAP

Modules in Undergraduate Mathematics and its Applications

Published in cooperation with the Society for Industrial and Applied Mathematics, the Mathematical Association of America, the National Council of Teachers of Mathematics, the American Mathematical Association of Two-Year Colleges, The Institute of Management Sciences, and the American Statistical Association.

Module 726

Calculus Optimization in Information Technology

Paul J. Campbell

Applications of Calculus to Computer Science

INTERMODULAR DESCRIPTION SHEET:	UMAP Unit 726
TITLE:	Calculus Optimization in Information Technology
AUTHOR:	Paul J. Campbell Mathematics and Computing Beloit College 700 College St. Beloit, WI 53511–5595 campbell@beloit.edu
MATHEMATICAL FIELD:	Calculus
APPLICATION FIELD:	Computer science
TARGET AUDIENCE:	Students in first- or second-semester calculus.
ABSTRACT:	Considers three optimization problems in the design of computer hardware and software and solves them by using differential calculus.
PREREQUISITES:	Students need to have studied optimization using derivatives to find and test critical points of a continuous function of one variable, and Section 4 uses the derivative of the natural logarithm function. Exercises 10 and 11 (which can be omitted) in Section 4 ask the student to calculate simple Taylor series to justify approximations used in the text. The exercises require a scientific-functions calculator.

©Copyright 1992 by COMAP, Inc. All rights reserved.

COMAP, Inc., Suite 210, 57 Bedford Street, Lexington, MA 02173
(800) 77–COMAP = (800) 772–6627,   (617) 862–7878

Calculus Optimization in Information Technology

Paul J. Campbell
Mathematics and Computing
Beloit College
700 College St.
Beloit, WI 53511–5595

Table of Contents

1. INTRODUCTION 1
2. MAXIMIZING STORAGE ON A DISK 1
   2.1 The Setting 1
   2.2 The Problem 2
   2.3 The Model 2
   2.4 The Optimum, According to the Model 3
   2.5 Comparison with Reality 3
3. OPTIMIZING DYNAMIC STORAGE 5
   3.1 The Setting 5
   3.2 The Problem 5
   3.3 The Model 5
   3.4 The Optimum, According to the Model 7
   3.5 Further Modeling and Analysis 7
   3.6 Comparison with Reality 7
4. MAXIMIZING THROUGHPUT ON A NOISY CHANNEL 8
   4.1 The Setting 8
   4.2 The Problem 9
   4.3 The Model 10
   4.4 The Optimum, According to the Model 12
   4.5 Comparison with Reality 12
   4.6 Sensitivity Analysis 14
   4.7 Comparison with Another Protocol 16
5. SOLUTIONS TO THE EXERCISES 18
6. APPENDIX: TECHNICAL DETAILS FOR COMPUTER SCIENCE STUDENTS 21
   6.1 Time vs. Space 21
   6.2 Current Disk Storage Technology 21
   6.3 The Arithmetic of Disk Storage 22
   6.4 Details of a Disk Track 23
   6.5 Error Correction 24
   REFERENCES 24
   ACKNOWLEDGMENTS 26
   ABOUT THE AUTHOR 26

MODULES AND MONOGRAPHS IN UNDERGRADUATE
MATHEMATICS AND ITS APPLICATIONS (UMAP) PROJECT

The goal of UMAP is to develop, through a community of users and developers, a system of instructional modules in undergraduate mathematics and its applications, to supplement existing courses and from which complete courses may be built.

The Project was initially funded by a grant from the National Science Foundation and has been guided by a National Advisory Board of mathematicians, scientists, and educators. UMAP is now supported by the Consortium for Mathematics and Its Applications (COMAP), Inc., a non-profit corporation engaged in research and development in mathematics education.

COMAP Staff

Paul J. Campbell	Editor
Solomon Garfunkel	Executive Director, COMAP
Laurie W. Aragón	Development Director
Philip A. McGaw	Production Manager
Roland Cheyney	Project Manager
Laurie M. Holbrook	Copy Editor
Dale Horn	Design Assistant
Rob Altomonte	Distribution Coordinator
Sharon McNulty	Executive Assistant

1. Introduction

In this Module, we will consider three optimization problems in the design of computer hardware and software:

- How can a computer disk be formatted to store the maximum amount of data?
- How can data files best be stored in blocks so as to minimize wasted space?
- How can we minimize time for transferring data between computers, while still checking for accuracy?

We will study the setting of each problem, construct a mathematical model to solve the problem, and compare the results of the theoretical solution to what is actually done. A common theme is that, although the problems are *discrete* (the solutions need to be integers), our models will be *continuous* (allowing real numbers as solutions) so that we can use the techniques of differential calculus.

2. Maximizing Storage on a Disk

2.1 The Setting

As manufactured, computer disks, both floppy diskettes and the platters in hard-disk drives, are just flat surfaces coated uniformly with a magnetic medium. To be used in a particular computer, they must be *formatted* in accordance with that machine's disk operating system. The formatting places little magnetic markers on the disk to divide it into fixed numbers of *sectors* and *tracks*. A sector is analogous to a sector of a circle, and a track is a thin circular ring (see **Figure 1**). Where the markers go, and how many sectors and blocks there are, depend on the kind of computer that the disk is to be used with. The markers are used by the operating system to find locations for storing and retrieving data on the disk. The formatting information, including spacing between the sectors, typically takes up 20–30% of the unformatted capacity.

The portion of a track within a single sector is called a *block*; a block is further subdivided into *bytes*, with each byte made up of eight *bits*. Each bit is a single small region that is magnetized or not, according to whether it is represents a 1 or a 0.

The number of tracks per inch is limited by mechanical considerations (how consistently accurately the disk controller can position the read/write head of the disk drive), while the maximum density of bits along a track is limited by magnetic considerations (the need to be able to distinguish two

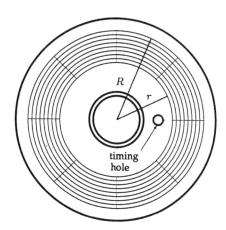

Figure 1. Schematic diagram of organization of data on a disk for which each track has the same number of sectors.

adjacent bits). The density of bits along a track is considerably greater than the density of tracks across the surface of the disk.

For design reasons, *every block on a disk contains exactly the same number of bytes*. For a disk like the one pictured in **Figure 1**, on which every track has the same number of sectors, the data bits along an outer track are farther apart than those on inner tracks.

2.2 The Problem

A key question is: For the disk to hold as much data as possible, where should the innermost track be located? We solve this problem by building a continuous model and then applying differential calculus.

2.3 The Model

Suppose the maximum feasible density for tracks is ρ_t tracks per inch (tpi) and the maximum density of bits along a track is ρ_b. Let the innermost track be at distance r (in inches) from the center of the disk, and let the outermost data track be at (fixed) radius R. Then the number of tracks is $(R-r)/\rho_t$.

The innermost track (and hence every track) contains $2\pi r/\rho_b$ bits. Then the total number of bits that the disk can store is

$$
\begin{aligned}
B(r) &= \text{(number of tracks)} \times \text{(number of bits per track)} \\
&= \frac{(R-r)}{\rho_t} \times \frac{2\pi r}{\rho_b} \\
&= Cr(R-r),
\end{aligned}
$$

where C is the constant $2\pi/\rho_t\rho_b$.

2.4 The Optimum, According to the Model

Using calculus, we find that the maximum of $B(r)$ occurs at $r^* = R/2$ (Exercise 1).

2.5 Comparison with Reality

To see how well theory agrees with reality, you can either make measurements on microcomputer diskettes (at the cost of destroying a diskette by opening it up) or else consult the standards for disk formats.

- **High-density double-sided 5.25-inch diskette.** This diskette holds 1.2MB (megabytes) when formatted for use in an IBM AT-compatible computer. This format has 96 tracks per inch, with track width 0.0061 inch. Only 80 tracks, numbered from 0 for the outermost track to 79 for the innermost, are used for data storage. The nominal location of the nth track is at a distance r_n (in inches) from the center of the diskette, with

$$r_n = x - \frac{n}{96}, \qquad n = 0, \ldots, 79,$$

and $x = 2.25$ for side 0 and $x = 2.167$ for side 1 (the two sides are offset by four tracks). For our purposes, we can regard the width of a track as negligible (in fact, it is 0.0061 inches).

Thus the effective R for this format is 2.25 inches for side 0, with the innermost track occurring at $r_{79} = 2.25 - 79/96 = 1.43$ inches; for side 1, the numbers are 2.167 and 1.34 inches. The data are written in a band between 1.34 and 2.25 inches from the center, which leaves an inner margin (from the outer edge of the small timing hole, 1.06 inches from the center) roughly equal to the outer margin (at the outer edge of the mylar, 2.56 inches).

There are 15 sectors per track and 512 bytes per sector, for $15 \times 512 = 7{,}680$ bytes per track. With 80 tracks, there are $80 \times 7{,}680 = 614{,}400$ bytes per side, or 1,228,800, or exactly 1,200KB (kilobytes), where 1K = 1,024.

The innermost track is track 79 on side 1, at 1.34 inches from the center. This track has length $2\pi(1.34) = 8.42$ inches, so the average density of user-databits along the innermost track is 60K/8.42 inches = 7,300 bits per inch, or 912—almost 1K—bytes per inch. (We are disregarding here the additional bits of formatting information that are also written on the track.)

The ratio of $r_0 = 1.34$ to the effective $R = 2.25$ is 0.60, not the one-half that our model predicts as optimum. With narrower margins, this diskette could hold more data (see **Exercise 2**).

- **Double-sided, double-density 3.5-inch diskette.** This kind of diskette is often labelled "1.0MB," referring to its unformatted capacity. Formatted for use in an IBM AT-compatible computer, it holds 720KB of user data. The format has 135 tracks per inch, with 512 bytes/sector and 9 sectors per side. The mylar of the diskette is 3.375 inches across, so $R \leq 1.68$ inch. The portion of the mylar with magnetic coating extends from the outer edge inward for a distance of 29/32 inch, to about 25/32 inch from the center; so $r \geq 0.78$ inch. The data lie on 80 tracks, which occupy about 0.60 radial inches. Allowing for equal inner and outer margins of 0.15 inch gives $r = 0.93$ and $R = 1.53$, with a ratio of 0.61. The innermost track holds 32K bits over $2\pi(0.93) = 5.84$ inches, for an average user-data density of 5,600 bits/inch, or 700 bytes per inch.

- **800K Macintosh format for 3.5-inch diskette.** The diskette drive built into Apple Macintoshes from 1985 to 1989 uses the same 3.5-inch diskette, with 80 tracks and 512 data bytes per sector. It puts more data on the diskette by varying the number of sectors per track [Chernicoff 1985, 379–380]:

  > ... [T]he 80 tracks are divided into five groups of 16 each, with the outermost group (tracks 0–15) holding 12 sectors per track, the next group (16–31) eleven sectors, and so on to only eight sectors each for tracks 64–79, the innermost group. This makes an average of ten sectors per track over the entire disk.... To keep the bit transfer rate constant, the disk drive runs its motor at any of five different speeds, depending on which group of tracks it is addressing; this is why the Macintosh "sings" when the disk drive is running.

  This approach offers a different kind of solution to our original optimization problem!

Exercises

1. Verify that $r^* = R/2$ is an absolute minimum of B:
   a) Show that r^* is a critical point of B.
   b) Check that this point is a relative maximum.
   c) Note the domain of B, check any endpoints, and conclude that r^* is an absolute maximum.

2. We have seen that a high-density double-sided 5.25-inch diskette has its innermost track at a distance about 60% of the effective R, instead of the $R/2$ that our model predicts as optimal. How much difference is there in data capacity? Can you make a general observation?

3. A formatted high-density double-sided 5.25-inch diskette holds 1.2MB of user data. With ρ_t, and ρ_b as given in the text, but data written right out to the edge of the mylar ($R = 2.56$), what would be the maximum user data capacity?

4. Suppose that the inner and outer margins on the high-density double-sided 5.25-inch diskette cannot be reduced. Could the diskette hold more data if the innermost track were farther from the center? Can you make a general statement?

3. Optimizing Dynamic Storage

3.1 The Setting

Both the main memory of a computer and its secondary storage (usually a disk) are often called upon to store data records and files whose lengths are not available in advance by the operating system. Ideally, the data should be stored in contiguous memory locations, to enable fastest retrieval of the data. However, a large enough block of contiguous memory may not be available.

One conventional solution to this problem is to store the data in a linked list, each of whose nodes has associated to it some fixed amount of storage (a "cell" or "block"), which is the same for each node. Part of the block is used to keep track of the list structure ("control information"), including the address of the next block in the list, and possibly additional information. The rest of a block is available for the data. The data file is broken into chunks, each of which—except possibly the last—fits exactly into the data area of a block; and the data file is stored in blocks corresponding to successive nodes of the linked list (see **Figure 2**).

Any unused part of the last block is wasted: the larger the block size, the greater potential waste. But if the block size is quite small, then a disproportionate amount of storage is taken up ("wasted") by control information. We will need to use the same size block for all files. Even so, the amount of wasted space will vary from file to file. We can minimize total wasted space in memory by minimizing the average amount of wasted space per file.

3.2 The Problem

What is the optimal size for a file storage block, to minimize the average amount of wasted space per file?

3.3 The Model

Following the notation of Wolman [1965], let us take c to be the size of the data portion of a block and b the size of the control information in each block.

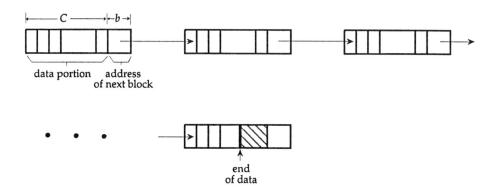

Figure 2. Storage of data in a linked list of "blocks."

Thinking of our problem as one of main memory storage, we may take c and b to be in units of machine words (a machine *word* is the smallest addressable unit of storage). As in our other models, we will want to consider c and b as continuous quantities, so that we can use calculus techniques in our analysis.

Suppose that on average our data records are L units long, with $b \ll L$ (b small compared to L). Then we will need

$$\left\lceil \frac{L}{c} \right\rceil$$

blocks, where the brackets denote the *ceiling function*: $\lceil x \rceil$ = the least integer not less than x.

A natural assumption is that the average amount of waste in the last block of the data storage is $c/2$.[1] In addition, the room for control information in that block is also wasted.

Then the average number of blocks that a record of length L will occupy is just

$$\frac{L}{c} + \frac{1}{2}.$$

Reasoning from **Figure 2**, the average amount of wasted space per record is given approximately as

$$W(c) = \frac{c}{2} + b\left(\frac{L}{c} + \frac{1}{2} + 1\right),$$

where the final 1 arises from needing to store control information in the directory table that indexes the contents of the entire memory.

---

[1]This assumption is equivalent to the data record lengths being evenly distributed over the congruence classes modulo $(c - b)$ [Trivedi 1982b, 74].

We find that
$$W(c) = \frac{c}{2} + \frac{bL}{c} + \frac{3b}{2}.$$

3.4 The Optimum, According to the Model

Using calculus, we find that the minimum of W occurs at $c^* = \sqrt{2bL}$ (**Exercise 5**).

3.5 Further Modeling and Analysis

Wolman [1965, 57–59] offers a more elaborate probabilistic model for the situation of longer data records being less likely than shorter ones. Specifically, he assumes that the length of a data record follows an exponential distribution, so that the *probability density function* for the length l of a data record is
$$f(l) = \frac{1}{L} \exp\left(\frac{-l}{L}\right), \qquad l \geq 0,$$
where L is the mean length of data records.

For this model, Wolman finds that
$$c^* = \sqrt{2bL} - \frac{b}{3} + \frac{b^{3/2}}{9\sqrt{2L}} - \frac{2b^2}{135L} + \cdots.$$

As Wolman notes, even for conservative values of b and L (such as $b = 2$, $L = 4$), the third and fourth terms offer negligible contributions. Since $b \ll L$, the optimal value of c is virtually unchanged from that for the simple model.

Wolman also develops discrete models, one using the *geometric distribution* for data record lengths and another assuming a uniform distribution of record lengths from 1 through some maximum R. For these models, too, the optimum value for c agrees closely with the value $\sqrt{2bL}$ determined for the simple continuous model.

Wolman also offers data to support the various models [1965, 67–68].

However, using a block size of $\sqrt{2bL}$ in connection with peculiar distributions of data record length (e.g., all data records of the same length, unfortunately incommensurate with the block size) can produce a non-minimal amount of waste (**Exercise 7**).

3.6 Comparison with Reality

In the case of disk storage, a record is a file and a block is a *cluster* of sectors. Diskettes and hard-disk drives formatted under PC-DOS/MS-DOS keep track of these clusters in a file allocation table (FAT). On a double-sided double-density 720K diskette, each cluster consists of two 512-byte sectors and has a 12-bit entry in the FAT. In units of bytes, we have $c = 1,024$ and $b = 1.5$.

Since the average length of a file stored on a diskette is probably 5K or 10K bytes, less waste would be produced with a smaller value for c. However, $c = 1,024$ bytes is dictated by other considerations; and b cannot be any smaller than 12 bits.[2] Moreover, a smaller block size would mean that for a fixed file length, more blocks would need to be located and read by the disk drive, thus likely increasing access time.

The size of the FAT (two copies of which are kept on the diskette) is fixed (3 sectors, or 1,536 bytes, per copy, for a 720K diskette), so that any of it not used is wasted anyway. Thus, the effective overhead is constant. **Exercise 8** asks you to estimate how inefficient the storage is on a 720K diskette.

Exercises

5. Verify that $c^* = \sqrt{2bL}$ is an absolute minimum of W on its domain.

6. Show that if b is small compared with L, then for the optimal value of c, the fraction of storage wasted is approximately $\sqrt{2b/L}$.

7. [Wolman 1965, 69] Investigate how optimal $c^* = \sqrt{2bL}$ is for the following situations, taking into account the fact that $(c + b)$ must be an integer number of words.
   a) $b = 0.5$ and every record has length 25, so $L = 25$;
   b) $b = 1$ and half the records have length 50 and half have length 150, so that $L = 100$.

8. How inefficient is the storage on a 720K PC diskette (which actually has 713K of user-data capacity)? Assume that $L = 10K$ bytes.

4. Maximizing Throughput on a Noisy Channel

4.1 The Setting

Communication within a computer, and between computers, must in many cases be perfect to be effective; if a copy of a program has even one bit miscopied, the copy may not run at all.

Some computer memories feature an extra *check* (or *parity*) *bit* for each eight-bit byte; the check bit is set to 1 if there is an odd number of 1s in the byte, to 0 if there is an even number. When the computer processes a byte, it also checks the check bit. If the check bit is not what it should be, then the data—either the check bit itself, or more likely one or more of the bits of the

---

[2]For hard drives with a capacity of 10MB or more, the FAT entry for a cluster has 16 bits, to accommodate the potentially larger number of files.

byte; we don't know exactly where the defect is—have become corrupted, and the data in the byte should not be used.

We will concentrate here, though, on communication by modem between computers. The bits of a file being transmitted are converted by the modem to audible tones (one pitch for a 1, another for a 0). Such communication often takes place over ordinary voice-grade long-distance telephone lines and is subject to "noise."

One way to check whether the data arrive as sent is for the communication to take place under a "protocol" that features some form of error-detection through calculation of a "check" quantity by the receiving computer. If an error is detected, then the part of the file that is affected must be re-sent. How complicated we make the check quantity depends on how sure we want to be to detect any errors. A single check bit will suffice to detect if a single error (or an odd number of errors) has occurred but will miss any even number of errors.

Checking and re-sending both slow down *throughput*, the volume of data that can be sent per unit time. If we were to wait until the end of the transmission to do a check, we might find that we have to re-transmit the entire file (and conceivably have to repeat it more than once!). On the other hand, sending a check bit with every byte of the file will slow down the rate of transmission of the file in two ways: we have nine-eighths as much to transmit, and the receiving computer must spend time checking each byte and signalling back whether it checks out.

So, there is a tradeoff between checking often and re-sending little vs. checking rarely and re-sending more (some of which was actually transmitted correctly—we just don't know which bits). Where to position ourselves on this tradeoff depends on how often data are corrupted in transmission—how noisy the line is.

We will build a model of the situation (using a little probability), analyze the model using differential calculus, and compare the results with some real data.

4.2 The Problem

Our goal will be to minimize the average total amount of time needed to send a file, including all re-transmissions of parts in which errors are detected. We will send the data in packets of b bits. After each packet is sent, a check quantity is sent; if an error is detected, the packet is re-sent. We want to determine the optimal size for b.

Note that since computer memories are designed to transfer data in *bytes*, (where a byte consists of eight bits), the lengths of both the packet and the check quantity will have to be whole numbers of bytes. However, considering that we are building a continuous model, it will do no harm for us to do most of our calculations in bits.

4.3 The Model

We assume that the original file consists of N bits, hence is sent as N/b packets (actually, $\lceil N/b \rceil$, the least integer not less than N/b; but remember, we are building a continuous model).

Let
$$p = \text{probability that a bit is received correctly.}$$

If we assume that errors are statistically independent, then the probability that an entire packet is received correctly is p^b, and the probability of a "bad packet" is $1 - p^b$. A result from probability theory[3] says that the average number of packets that need to be sent to achieve N/b successes is N/bp^b. So, on average, transmitting the file will take

$$\frac{N}{bp^b} \times \text{(time to send a packet, including the check quantity).}$$

We must also include in the time to send a packet the time to send control information (start-of-packet character, packet length, packet sequence number), send the check information, process the check information, and respond whether the check is OK or not. A simple model for the time needed is to assume that it does not depend on the length of the packet: that is, we use a single check digit or a small number of check digits, the same regardless of the length b of the packet. The drawback of so simple a scheme is that for a longer packet length, there is a greater chance of multiple errors that our check quantity will not detect. On the other hand, we would like to avoid using a longer and more complicated check quantity whose transmission would require a time proportional to the length of the packet.

With this assumption, the average time to complete a correct transmission of our message is

$$f(b) = \frac{N}{bp^b} k(b + \lambda),$$

where k is the time it takes to send one bit and λ is the number of bits in the check quantity. For fixed λ, we want to maximize $f(b)$ for values of b between 1 and N, inclusive. **Figure 3** shows graphs of f for a typical value $\lambda = 8$ and varying values of p.

We analyze f for its extreme values by finding its derivative (**Exercise 9**):

$$f'(b) = \frac{-Nk}{b^2 p^b} \left[(\ln p)b^2 + (\lambda \ln p)b + \lambda \right].$$

For the special case $p = 1$ (perfect transmission), we have $\ln p = 0$ and the derivative simplifies to $-Nk\lambda/b^2$, which is never zero; so the extreme values are at the endpoints ($b = N$ to minimize f, and $b = 1$ to maximize f).

---

[3] The mean of the negative binomial distribution for the the kth success, with $k = N/b$ and probability of success p^k, is N/bp^b [Larsen and Marx 1986, 222–224].

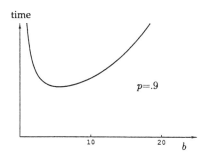
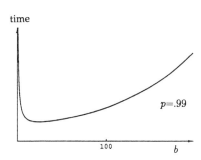
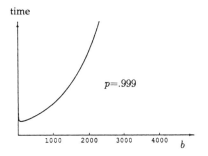
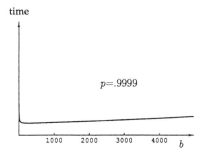

Figure 3. Graphs of $f(b)$ for $\lambda = 8$ and $p = .9, .99, .999, .9999, .99999$, and $.999999$.

Otherwise, by using the quadratic formula, we find that the derivative is zero at
$$b^* = \frac{-\lambda \ln p \pm \sqrt{(\lambda \ln p)^2 - 4\lambda \ln p}}{2 \ln p}.$$

This expression suggests substitution of the positive quantity
$$q = -\lambda \ln p,$$

as well as the use of $|\ln p| = -\ln p$ (since $p < 1$), to get
$$b^* = \frac{1}{|\ln p|} \frac{q}{2} \left(1 \pm \sqrt{1 + \frac{4}{q}}\right).$$

Now, for an efficient check quantity, λ will be fairly small, certainly no more than a few hundred bits at most. Under normal circumstances, p will be close to 1, so $-\ln p$ will be close to 0. Hence $q = -\lambda \ln p$ will be close to 0. In such circumstances (**Exercise 10**),
$$1 + \frac{4}{q} \approx \frac{4}{q}$$

and
$$1 \pm \sqrt{1 + \frac{4}{q}} \approx \pm \frac{2}{\sqrt{q}}.$$

We will be interested in only the positive result.

4.4 The Optimum, According to the Model

Substituting the approximation into the expression for b^*, we get
$$b^* \approx \frac{\sqrt{q}}{|\ln p|} \approx \sqrt{\frac{\lambda}{|\ln p|}}.$$

To make this formula meaningful, we can examine some numerical instances, as in **Table 1**. Note that for values of p near 1, we have $|\ln p| \approx 1 - p$ (**Exercise 11**).

4.5 Comparison with Reality

To compare our model with reality, we offer data from the use of the FORMAC protocol for transfers between an IBM mainframe and a Macintosh II [Simware 1989]. The packet size for this protocol is 1,905 bytes, or 8 × 1,905 = 15,240 bits, and it takes about 10 sec to transfer one such packet over a 2400-bps (bits per second) modem.

We transferred a file of length 105K bytes, which was sent by the protocol as 60 packets (including check information). Of the 60, 35 got through correctly the first time, about a 60% rate. For the remaining 25, we cannot

Table 1.

Results for a check function consisting of λ check bits. Optimal packet length b^* (in bits, rounded up to the next integer) and probability P of correct transmission of a packet, as functions of the probability p of correct transmission of a single bit.

λ	p	$1-p$	$b^* = \sqrt{\lambda/|\ln p|}$	$P = p^{b^*}$
1	.9	.1	3	.729
	.99	.01	10	.904
	.999	.001	32	.968
	.9999	.0001	100	.990
	.99999	.00001	316	.997
	.999999	.000001	1,000	.999
8	.9	.1	9	.387
	.99	.01	28	.755
	.999	.001	89	.915
	.9999	.0001	283	.972
	.99999	.00001	894	.991
	.999999	.000001	2,828	.997
16	.9	.1	12	.282
	.99	.01	40	.669
	.999	.001	126	.882
	.9999	.0001	400	.961
	.99999	.00001	1,265	.987
	.999999	.000001	4,000	.996

really be sure how many times a packet was re-sent. The message "bad block [packet]—attempting re-transmission" blinked at 10-second intervals, which could indicate either another unsuccessful attempt or else that the mainframe operating system had not yet undertaken the task. In any case, the delay incurred, whether from re-sending or waiting to re-send, reflects the time cost of the original (and any subsequent) error. We recorded that 15 of the 25 erroneous packets were "re-sent" once, 6 were "re-sent" twice, 1 was "re-sent" 6 times, 1 was "re-sent" 15 times, 1 was "re-sent" 19 times, and 1 was "re-sent" 23 times. The entire transmission took about 23 min.

Since about 60% of the transmissions got through the first time, we would expect about 60% of the first retries to get through, and 15 of 25 did. We would expect about 60% of the second retries to get through, and 6 of 10 did. But we wouldn't expect needing up to 23 re-transmissions! One explanation may be that the mainframe is heavily burdened at certain times, and transmission attempts at those times suffer long delays. Another explanation is that the communications line may be very noisy for certain intervals, so that errors are more likely and more frequent during those times; errors may occur in *bursts*, so that our assumption that errors are independent does not hold. The result cited above from probability theory shows that, without any overburdening of the mainframe or any noisy intervals, it should have taken on average about $N/bp^b = (N/b)/p^b = 60/0.60 = 100$ packet attempts, or 17 min, to transmit the file.

The probability $P = p^b$ of successful transmission of a packet is about 60%; with $b = 15{,}240$, we arrive at the estimate $p = .999966$. (**Exercise 12** asks you to tell why this value does not contain an excessive number of decimal places.) For such a value of p, the optimal packet length is $b^* = \sqrt{\lambda/|\ln p|} \approx 170\sqrt{\lambda}$. We do not know the value of λ that is used by the FORMAC protocol[4], but we can evaluate the expression for b^* with different possible values of λ. In particular, for $\lambda = 8$ (the common situation of a check *byte*), we get $b^* = 480$; for $\lambda = 1{,}024$—a huge amount of checking!—we get $b^* = 5{,}400$. Both of these values are far short of the block size used by the protocol, 15,240.

Another way to interpret the block size used by the protocol is to ask an "inverse" question: For what value of p is the protocol's block size the optimal block size? (**Exercise 13**).

4.6 Sensitivity Analysis

The difference between the optimal packet size and the packet size used by the protocol raises a question relevant in most real-world modeling situations: Just how sharp is the optimum? How much does it matter if we use a packet that is twice as long, or half as long, as the optimal size? This is certainly a concern here, as there is no way of knowing in advance how noisy the

---

[4]There is no mention of it in the manual [Simware 1989], and the vendor did not respond to a letter requesting more information.

communications channel will prove to be, that is, the value of p. If the user can't adaptively change the packet size, the author of an error-checking protocol needs to cover a wide range of possible values of p that may occur.

In fact, the minimum is fairly broad, as you can see from **Figure 4**. The figure is for $p = .999966$, but the graph is similar for any value of p close to 1. **Exercise 13** asks you to do some relevant calculations, which show that— under the circumstances of the experiment—the large block size used by the protocol seems to result in substantial inefficiency.

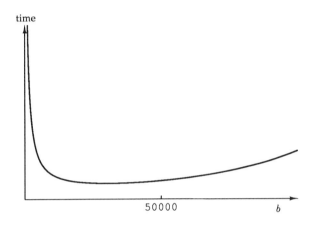

Figure 4. Graph of $f(b)$, with $\lambda = 8$, for $p = .999966$.

Further use of calculus allows us to investigate analytically how broad the minimum is. Putting the matter another way, with calculus we can assess the sensitivity of the function $f(b)$ to perturbations of $b = b^* + \Delta b$ from the optimum value b^*. For many scientific purposes, it makes sense to think in terms of *relative* change away from b^*—as in, say, a change by a factor of 2—so we write Δb as a fraction of b^*, i.e., $\Delta b = \epsilon b^*$. We calculate $f''(b^*)$ and expand f in a Taylor series around $b = b^*$. After some calculation (**Exercise 14**), we find

$$f(b^* + \epsilon b^*) = f(b^*) \left(1 + \frac{\epsilon^2}{2} (1 + |\ln p|) \sqrt{\lambda |\ln p|}\right).$$

For p close to 1, we have $(1 + \ln p) \approx 1$, so

$$\frac{f(b^* + \epsilon b^*)}{f(b^*)} = \left(1 + \frac{\epsilon^2}{2} \sqrt{\lambda |\ln p|}\right).$$

Now, for most protocols, λ will not be very large (a few bytes, at most), while $\ln p \approx 1 - p$ will be quite small, usually less than 0.01. Their product will be less than 1. Thus, the relative change in $f(b)$ will be less than $\epsilon^2/2$.

For $p = .99$ and $\lambda = 8$ (a single check byte), we explore changes in b^* by a factor of 2 in each direction:

If b is only half the optimal value ($\epsilon = -0.5$), then the transmission time is about 3% longer than if we were at the optimum.

If b is twice the optimal value ($\epsilon = 1$), the time is about 14% longer.

Note that for the data in our experiment, we are close to the first of these situations.

We dare not extrapolate for values of ϵ close to -1! How far we can extrapolate validly can be determined by exploring formulas for the Taylor series remainder, or by considering the cubic term of the Taylor series. We are approximating the curve by a parabola; glancing at the curve itself—or better yet, using a computer to graph both the curve and the approximation—can give good clues about where the approximation starts to be useless for our purposes.

How should the designer of a protocol choose the packet length? Our investigation tells us that it is better for the packet length to be one-half of the optimal length than to be twice it; but we haven't explored more extreme cases, such as a factor of 10.

4.7 Comparison with Another Protocol

Some communications protocols allow the user to customize the protocol. The well-known public-domain Kermit protocol [da Cruz and Catchings 1984] offers a choice of packet size and check methods. The default packet size is 90 bytes (= 720 bits); there are 4 bytes of control information; and the user may choose to use 1, 2, or 3 check bytes (for $\lambda = 8$, 16, or 24), corresponding to different check schemes.

With a default packet length of 720 and $\lambda = 8$, and assuming other factors equal, how would we expect Kermit to compare with FORMAC (also with $\lambda = 8$) under the conditions of the FORMAC experiment (where $p = .999966$)? We find

$$\frac{f(720)}{f(15240)} = 0.62.$$

So, transmission of the 105 KB file using Kermit should take only 62% of the 23 min required with FORMAC, or about 14 min. In fact, our tests show that transmission using Kermit takes a little over 17 min, the difference reflecting overhead. Recall that we predicted that FORMAC should take only about 17 min, if it weren't for intervals when the mainframe was busy or the line was excessively noisy (conditions that violate our model's assumptions).[5]

---

[5] A modem has to transmit not only user data but also the bits for any error-checking, flow control, and hardware "handshaking." Without any of this overhead, transmitting the 105 KB on a 2400-bps modem would take 5.8 min.

Exercises

9. Verify the calculation of $f'(b)$.

10. Use calculus to show that $\sqrt{1+x} \approx 1 + x/2$ for x close to 0.

11. Use calculus to show that $|\ln p| \approx 1 - p$ for p close to 1.

12. We calculated $p = .999966$ from $b = 15,240$ and an estimate that $P = .60$. Explain why our value of p has the correct number of significant digits.

13. Assuming that the FORMAC protocol uses a single check byte, for what value of p is a block size of 15,240 bits the optimal block size?

14. Verify the calculation of $f(b^* + \epsilon b^*)$.

15. With $p = .999966$ and $\lambda = 8$, we found that the optimal packet size is $b^* = 480$. Investigate the sharpness of this optimum by computing the ratio $f(b)/f(b^*)$ for the following values of b:
    a) $b = 256$ (about half the optimal size),
    b) $b = 1,024$ (about twice the optimal size),
    c) $b = 15,240$ (the size used in the protocol).

16. Perform a sensitivity analysis on c^* in **Section 3.4** similar to the one done for b^* in this section.

17. [da Cruz and Catchings 1984, 400–401] In addition to the data (with block length b), a Kermit packet contains 5 bytes of control information and 1 to 3 bytes of check information. Also, 5 bytes are returned in the acknowledgement packet from the receiving machine. However, since control characters (that is, characters formed by using the CTRL key plus another key) are used to control the data communication, any control character that is part of the data must be replaced with a prefix plus the character. Even simple text contains carriage return (and sometimes line feed) control characters for each line; we may suppose that adding prefixes for these and other control characters in text adds an overhead of about 5% to the data. We include in the overhead the bytes of check information but not what is involved in initiating and terminating the file transfer, since that is fixed regardless of block length. With this information, show that
    a) Kermit with a block size of 40 bytes spends about 25% to 27% of the time in overhead, depending on how many bytes of check information are used;
    b) Kermit with a block size of 96 bytes spends about 14% to 16% of the time in overhead.

5. Solutions to the Exercises

1. **a)** $B'(r) = C(R - 2r)$.
   b) Either note that $B''(r) = -2C$, which is negative for $r = r^*$, and cite the second derivative test; or check that $B'(r)$ is positive to the left and negative to the right of of r^*, and cite the first derivative test; or observe that $B(r*) = CR^2/4 > 0$ and that $B(r)$ is 0 at the endpoints 0 and R of the interval. You can do this part without calculus if you observe that the function's graph is a parabola symmetric around $R/2$.
   c) The domain of B is $0 \leq r \leq R$, and $B(0) = B(R) = 0 < B(r^*) = CR^2/4$.

2. $B(0.6R) = 0.24CR^2$, while $B(0.5R) = 0.25CR^2$, a difference of only 4%. We say that the minimum is *broad*.

3. With $r = 1.28$ inches (a value that would not interfere with the timing hole), one side could hold over 900KB, so the disk could hold 1,800KB. Of course, some margin from the edge is essential.

4. For the diskette in question, the domain of B is $0.6R \leq r \leq R$, over which B is decreasing. Hence the diskette could not hold more data if the innermost track were farther from the center.

5. **a)** We have
$$W'(r) = \frac{1}{2} - \frac{bL}{c^2}.$$
   b) $W''(r) = 2bL/c^3$, which is positive for $c = c^*$.
   c) The domain of W is $c \geq 1$, so the endpoint $c = 1$ should be checked.

6.
$$\begin{aligned}\frac{W(c^*)}{L} &= \frac{\sqrt{2bL}}{2L} + \frac{bL}{L\sqrt{2bL}} + \frac{3b}{2L} \\ &= \sqrt{\frac{b}{2L}} + \sqrt{\frac{b}{2L}} + \frac{3b}{2L} \\ &\approx 2\sqrt{\frac{b}{2L}} = \sqrt{\frac{2b}{L}}.\end{aligned}$$

7. **a)** The formula gives $c^* = 5$, which suggests that c should be one of the neighboring feasible discrete values, 4.5 or 5.5. Either of these values leads to using 30 words to store each record, whereas taking c to be either 12.5 or 25.5 requires only 26 words to store a record.
   b) The formula gives $c^* = 10$; a record of length 50 will require 55 words, and one of length 150 will require 165 words, for an average waste of 10 words per record. Taking c to be 50, instead, would require 51 words for a record of length 50 and 153 words for one of length 150, for an average waste of 2 words per record.

8. Suppose the diskette is just about full, with about 71 files and not enough room to store another file of average size. Using $b = 1.5$ and $L = 10K$, we get $c^* \approx 175$ bytes, $W(c^*) \approx 180$ bytes, and $W(1,024) \approx 530$ bytes per 10KB file, so that the waste is about 5%. In addition, the remaining diskette space, which is not large enough to store another file and on average would be about half the length of an average file (5K), is also wasted. In fact, there is a constant overhead (boot sector, directory, two copies of the FAT) of 7K for the diskette; on average, each file wastes half of its last cluster, or 512 bytes, for a total of 35.5K; and we lose half the length of an average file, 5K. Total waste: 47.5K, or about 7%. What about our assumption of an average file length of 10K? The directory on the diskette has room for at most 112 file entries, thus forcing a full diskette to have an average file length greater than 6K; a 1.44MB diskette, with twice the data capacity and twice the directory size, is in the same situation.

9.
$$\begin{aligned} f'(b) &= N\left(\frac{-p^b - bp^b \ln p}{(bp^b)^2}\right)k(b+\lambda) + \frac{N}{bp^b}(k) \\ &= Nk\left(\frac{bp^b - (b+\lambda)p^b(1 + b\ln p)}{(bp^b)^2}\right) \\ &= \frac{-Nk}{b^2 p^b}\left[(\ln p)b^2 + (\lambda \ln p)b + \lambda\right]. \end{aligned}$$

10. Expanding $\sqrt{1+x}$ as a Taylor series around $x = 0$, we get
$$\sqrt{1+x} = 1 + \frac{x}{1 \cdot 2!} - \frac{x^2}{4 \cdot 2!} + \cdots,$$
which yields the stated approximation.

11. Expanding $\ln(1+x)$ as a Taylor series around $x = 0$, we get $\ln(1+x) = x - \frac{x^2}{2} + \cdots$, leading to the approximation $\ln(1+x) \approx x$ being a good one for $x \approx 0$. We take $p = 1 + x$, or $x = p - 1$, leading to $\ln p \approx p - 1$.

12. Because P has only two significant digits, so should our estimate for p. In fact, it does. We calculate $\ln p = -0.0000034$ (with two significant digits), and we know (from **Exercise 10**) that for values of p near 1, we have $\ln p \approx p - 1$. The initial 9's in the expansion for p are just placeholders; like initial 0's, they do not count as significant digits.

13. Solve $b^* = \sqrt{\lambda/|\ln p|}$ for p, getting $p = \exp\{-\lambda/(b^*)^2\}$. For $\lambda = 8$ and $b^* = 15,240$, we get $p = .999999966$.

15. The ratios are 1.01, 1.01, and 1.61. While a packet size of half or double the optimal size increases the transmission time by less than 1%, the very large size used by the protocol increases the transmission time by more than 60%. The size used by the protocol is adapted to better transmission conditions, which favor a larger block size.

16.
$$W(c^* + \epsilon c^*) = \frac{3b}{2} + \frac{\sqrt{2bL}}{2}\left(1 + \epsilon^2\right).$$

We may assume that the first term is negligible compared to the first. The minimum is fairly sharp, in that using a block size twice as long as optimum ($\epsilon = 1$) leads to almost twice as much waste, while using one half the optimal size ($\epsilon = -0.5$) leads to almost a quarter more waste.

17. For each block of b bytes of data, Kermit must send an additional $A = 0.05b + 10 + c$ bytes, where c is the number (1 to 3) of check bytes. The proportion of overhead is thus $A/(b + A)$, and this formula yields the percentages claimed.

6. Appendix: Technical Details for Computer Science Students

6.1 Time vs. Space

We have considered in this Module optimizations of both time (for modem transmission) and space (for disk storage and dynamic storage), noting in each case that there were certain design constraints. Optimization of either time or space separately may be at the expense of the other; overall optimization of system performance, in light of economic considerations, usually involves a tradeoff between the two.

For example, McKusick et al. note that a file block size of 512, as used in a then-"traditional" UNIX system, was by 1984 "incapable of providing the data throughput [disk read and write] rates that many applications require" [1984, 182]. That block size afforded only 4% wasted disk space on their system. Changing to a block size of 4,096 as the minimum unit for a read or a write increased throughput by a factor of about 10 but—as we would expect—increased wasted space, to 46/some blocks allowed the authors to reduce wasted space to 7/8,192; but there wasn't much further improvement, and they had reached the point that the throughput was now limited by the speed of the CPU.

The interplay and balance of conflicting design considerations is the underlying theme in courses in operating systems.

6.2 Current Disk Storage Technology

For comparison to the figures that follow, we repeat here the data for two common diskettes:

- a 720 KB 3.5-inch IBM-format floppy diskette has a *track density* 135 tpi and a *track capacity* of about 5,600 bpi;

- a high-density 1.2 MB 5.25-inch diskette has 96 tpi and 7,300 bpi.

Track capacity is determined by the ability to store changes in magnetic flux and by the technique used to encode the data. Increasing the track capacity made it possible to change the capacity of a 3.5-inch diskette from 720KB to 1.44MB and then (using a different medium) to 2.88MB; the 2.88MB diskettes have a density of 35,000 bpi.

Track density is "stuck" at 135 tpi for *magnetic* media because the drive head cannot be positioned accurately enough without a change in technology [Floppy... 1992].

Drives current in 1992 store 60,000 bpi and have from 1,500 to 2,000 tracks per inch. Using the latter figure, the capacity can be stated as 2.5 MB/cm^2.

Experts expect the technology to reach 2,500–3,500 tpi by 1993 or 1994 [McGrath 1991, 260] and 120,000 bpi and greater [Ryan 1992, 167, 169].

Optical technology can offer more than 20MB of storage on a 3.5-inch optical diskette. The optical technology uses permanently marked "servo" tracks that the drive uses as reference points to position the head very accurately; these "embedded servos" can take up about 10% of the data surface [McGrath 1991, 260]. The PLI Floptical has a track density of 1,245 tpi [Coming of age ... 1992], while Brier Technology has announced a 50MB floppy [Floppy ... 1992].

Stein [1992] notes that the theoretical limit for storage on optical diskettes is 50 MB/cm^2, *assuming* that the illumination source has a width of 5×10^{-7} meters. Researchers at AT&T Bell Laboratories announced in August, 1992, though, that they had developed a magneto-optical system with a density of 5,600 MB/cm^2. The key to the new technology is funneling the laser light not through a lens but through an optical fiber stretched to one one-thousandth the thickness of a human hair, for a width of 5×10^{-8} meters. However, the laser is still too slow for commercial applications, particularly real-time playback of video [It's ... Superdisk 1992].

A necessary and sobering consideration, however, is that disk surface "defect densities grow exponentially with increases in track densities" [Kirk et al. 1992, 201]. Hence, researchers are exploring the possibility of using lasers to store data on individual molecules of dye embedded within tiny beads of polystyrene [Freedman 1992].

6.3 The Arithmetic of Disk Storage

Table 2 shows data on various diskettes and hard disk drives.

Table 2.
Specifications for IBM-formatted disks, with 512KB in each sector.

	360KB	1.2MB	720KB	1.44MB	20MB	40MB	300MB
tracks per inch	48	96	135	135			
platters	1	1	1	1	2	4	8
surfaces	2	2	2	2	4	8	16
tracks/surface	40	80	80	80	615	731	≥2,631
sectors/track	9	15	9	18	17	17	14
total sectors	720	2,400	1,440	2,880	41,820	99,416	
reserved sectors	12	29	14	33	2,524	5,976	
user sectors	708	2,371	1,426	2,847	39,296	93,440	588,672
user bytes	362,496	1,213,952	730,112	1,457,664	20,119,552	47,841,280	301,400,064
sectors/cluster	2	1	2	1	4	8	16

The numbers that the DOS command CHKDSK will show for total (user-available) disk space and size of allocation unit (cluster) will agree with the table for diskettes but may vary with manufacturer for hard drives.

For each file, the directory of an IBM disk stores a pointer to an entry in the file allocation table (FAT). The sectors used by the file are listed in the FAT. The smallest block of file storage is called a *cluster* or *allocation unit*; it varies in size with the disk, from a single sector to 8 or more sectors, in powers of 2.

The reserved sectors include one for the boot sector, plus room for the directory and for two copies of the FAT.

Versions of DOS prior to 4.0 number each sector on a hard disk with a 16-bit number, thereby allowing for up to $2^{16} = 65,536$ sectors, or 32MB. Even for larger disks, though, the FAT is limited in size to 65,535 entries (of 2 bytes each). As a result, a large hard disk *must* allocate storage in clusters larger than a single sector. For example, for a hard drive to hold more than 256 MB, it must use a cluster size of at least 16, or 8,192 bytes per cluster [DeVoney 1989, 449–450].

6.4 Details of a Disk Track

From 10% to 50% of the area of a disk is occupied by (necessary) spaces between the tracks, while either the inner or the outer margin is used as a "landing zone" area to store information on bad blocks.

Of the remaining potential storage area, or unformatted capacity, as much as 30% may be occupied by formatting information. This information includes necessary overhead for each track, consisting of pre-index gap, index address mark, and a post-index gap, followed by the sectors on the track, and then a final gap to finish the track. For each sector, there is

- a sector ID field ("header")—usually 7 bytes,

- a post-ID-field gap,

- the user data field,

- error-detecting or error-correcting check bytes for the user data, and

- a post-data-field gap.

The sizes of the gaps depend on disk-drive motor speed, motor speed variation, sector length, interleaving, and amount of error correction capability. For example, the the Seagate Elite 2 disk [Seagate 1992, 4] uses 82 bytes of overhead for each 512-byte user data sector.

6.5 Error Correction

The subject of error control in data transmission and storage is a broad area that requires facility in abstract algebra; but the applications are so numerous and ubiquitous that a computer science student should not neglect study of coding theory.

We have examined the costs of error control in data transmission without going into the specifics of the many coding methods that are used in different applications. Here we say a little about error correction methods in magnetic disks, for which there is less variation.

The sector header information typically includes an error-*detecting* code, usually the two-byte cyclic redundancy check code (CRC–16) based on the 16-degree polynomial $x^{16} + x^{12} + x^5 + 1$. For how such a code works, see Ralston and Reilly [1983, 434–437]; and for its capabilities, including protection against burst errors, see Intel Corporation [1985, 6–676].

For the user data, there may be either just error detection (again using CRC–16) or more-sophisticated error *correction* (using more bytes). For example, the Intel 82064 Winchester disk controller chip, with on-chip error detection and correction, can employ either CRC–16 error detection or else an error-correction code that takes up 4 bytes per sector and uses a 32-degree polynomial [Intel Corporation 1985, 6–640]. The Seagate Elite 2 uses 8 bytes per sector and a 64-degree polynomial [Seagate 1992, 4], while other disks use 11 bytes per sector and an 88-bit Reed-Solomon code [Kirk et al. 1992].

Many textbooks on coding theory treat it as a realm of pure mathematics, but there are numerous sources for the details of contemporary applications:

- "older" data storage systems (IBM System 370 memory, 6250 bpi magnetic tapes, IBM disk systems) [Lin and Costello 1983, 498–532];

- spacecraft communications [Lin and Costello 1983, 533–546; Hill 1986, 9–10; Welsh 1988, 67–74];

- telephone and submarine cable transmissions [Lin and Costello 1983, 547–550];

- packet radio [Lin and Costello 1983, 551–559]; modems [Calderbank 1991]; and

- audio compact discs [Hoffman et al. 1991, 182–184, 249–252].

References

Calderbank, A.R. 1991. The mathematics of modems. *Mathematical Intelligencer* 13 (3) (Summer 1991): 56–65.

Chernicoff, Stephen. 1985. *Macintosh Revealed. Vol. 2: Programming with the Toolbox*. Hasbrouck Heights, NJ: Hayden.

Coming of age .... 1992. *MacUser* 16 (9) (September 1992): 213.

da Cruz, Frank, and Bill Catchings. 1984. Kermit: A file transfer program for universities. Part 1: Design considerations and specifications. Part 2: States and transitions, heuristic rules, and examples. *Byte* 9 (6) (June 1984) 255–278; 9 (7) (July 1984): 143–145, 400–403.

DeVoney, Chris. 1989. *Using PC DOS*. 3rd ed. Carmel, IN: Que Corp.

Floppy—but very large. 1992. *Byte* 17 (3) (March 1992): 166–167.

Freedman, David H. 1992. Drawing a bead on superdense data storage. *Science* 255 (6 March 1992): 1213–1214.

Gillett, Philip. 1984. *Calculus and Analytic Geometry*. 2nd ed. Lexington, MA: D.C. Heath.

Hill, Raymond. 1986. *A First Course in Coding Theory*. New York: Oxford University Press.

Hoffman, D.G., D.A. Leonard, C.C. Lindner, K.T. Phelps, C.A. Rodger, and J.R. Wall. 1991. *Coding Theory: The Essentials*. New York: Dekker.

Intel Corporation. 1985. *Microsystem Components Handbook 1986. Peripherals. Vol. II*. Santa Clara, CA: Intel Corporation.

It's ... Superdisk. *Time* 140 (7) (17 August 1992): 18.

Kirk, Rod, Tim Christianson, and Danial Faizullabhoy. 1992. Embedded intelligence. *Byte* 17 (3) (March 1992): 195–203.

Larsen, Richard J., and Morris L. Marx. 1986. *An Introduction to Mathematical Statistics and Its Applications*. 2nd ed. Englewood Cliffs, NJ: Prentice-Hall.

Lin, Shu, and Daniel J. Costello, Jr. 1983. *Error Control Coding: Fundamentals and Applications*. Englewood Cliffs, NJ: Prentice-Hall.

McGrath, James. 1991. The incredible shrinking disk. *Byte* (October 1991): 255–264.

McKusick, Marshall K., William N. Joy, Samuel J. Leffler, and Robert S. Fabry. 1984. A fast file system for UNIX. *ACM Transactions on Computer Systems* 2 (3) (August 1984): 181–197.

Ralston, Anthony, and Edwin D. Reilly, Jr. 1983. *Encyclopedia of Computer Science and Engineering*. 2nd ed. New York: Van Nostrand Reinhold.

Ryan, Bob. 1992. Scaling the memory pyramid. *Byte* 17 (3) (March 1992): 161–170.

Seagate Corporation. 1992. Engineering Specification: Media and Format Specifications for the Elite 2 Disc Drive Model Number ST42400N and ST42400ND. Spec. 64803300, Rev. A, 1-9-92.

Simware. 1989. *Manual for SIM3278*. Ottawa, Ontario: Simware, Inc.

Stein, Richard Marlon. 1992. Terabyte memories with the speed of light. *Byte* 17 (3) (March 1992): 168–169.

Trivedi, Kishor S. 1982a. *Probability and Statistics with Reliability, Queuing, and Computer Science Applications*. Englewood Cliffs, NJ: Prentice-Hall.

Trivedi, Kishor S. 1982b. *Solutions Manual: Probability and Statistics with Reliability, Queuing, and Computer Science Applications*. Englewood Cliffs, NJ: Prentice-Hall.

Welsh, Dominic. 1988. *Codes and Cryptography*. New York: Oxford University Press.

Wolman, Eric. 1965. A fixed optimum cell-size for records of various lengths. *Journal of the Association for Computing Machinery* 12: 53–70.

Acknowledgments

The section on maximizing storage on a disk was inspired by Gillett [1984, 217–218]. I am indebted to Mark Rogers, technical support engineer for Verbatim, for pointing me to the specific ANSI standard (X3.162–1988) for high-density 5.25-inch disks.

The section on dynamic storage arose as an elaboration of a problem in Trivedi [1982a, 187, #1], who cites Wolman [1965]. Wolman attributes the simplified model and its solution to J.B. Kruskal.

I thank Eric Bach (Computer Sciences Dept., University of Wisconsin—Madison), Chris Carroll (Dept. of Computer Engineering, University of Minnesota—Duluth), and Darrah Chavey (Dept. of Mathematics and Computer Science, Beloit College) for offering helpful comments on portions of this Module. Michael Anctil of Magnetic Peripherals supplied information on Seagate products in development.

An editor who takes on the role of author hazards being less objective and perceptive about the quality of the writing. Both author and readers can be served best by another party taking a hand in editing the work. For his help in that regard, I am grateful to Phil Straffin of Beloit College.

This Module was developed in connection with the NSF Calculus Reform in the Liberal Arts College Project, supported by NSF grant USE 8813914. An abbreviated version will appear in the volume *Applications of Calculus*, edited by Philip D. Straffin, to be published by the Mathematical Association of America in 1992.

About the Author

Paul Campbell is professor of mathematics and computer science at Beloit College, where he served as Director of Academic Computing from 1987 to 1990. He has been the editor of *The UMAP Journal* since 1984.